Praise for *Catastrophism*

"Here you have it, a forceful rejection of that gleeful, adolescent paranoia that masquerades as hardcore realism. As the essays in this excellent book show, 'catastrophism' is a *wish* expressed as a *fear*, a masochistic cop-out that postures as bravery. Welcoming the end of the world as the catalyst of political deliverance is one of the most irresponsible positions on offer. This book is a superb antidote to the unproductive politics of fear."
—Christian Parenti, author of *Tropic of Chaos: Climate Change and the New Geography of Violence*

"*Catastrophism* comes at the right moment: the year of The End proclaimed across the political spectrum from deep ecologists to the Mayan Calendarists. Instead of concentrating on the merits of the claims of the various apocalypticians, James Davis, Sasha Lilley, David McNally, and Eddie Yuen examine the political function of these claims and find them to be deeply reactionary. This is a controversial book that challenges many of the unexamined assumptions on the left (as well as on the right). It is a warning not to abandon everyday anticapitalist politics for a politics of absolute fear that inevitably leads to inaction."
—Silvia Federici, author of *Revolution at Point Zero: Housework, Reproduction, and Feminist Struggle*

"Bravo! This is the book that has been sorely needed for so long to reveal the dead end that a politics founded on catastrophic predictions must lead to in terms of either preventing them or actually changing the world. Essential reading for all those on the left who are concerned with the question of strategy [illegible]
—Leo Panitch, coauthor of *The Making of Global* [illegible] *In and Out of Crisis*

"At last, and not before time, a full-spectrum guidebook to catastrophism. Together Lilley, Yuen, Davis, and McNally penetrate the smoke of apocalypse, past the politics of fear and redemption which is the stock-in-trade of disaster merchants—right, left, and green. If we and our fellow species are to leave the dark shadow of John the Divine and Parson Malthus, this superb, clear-eyed collaboration opens the way."
—Iain Boal, coeditor of *West of Eden: Communes and Utopia in Northern California*

"I cannot overstate how critically important this volume is. *Catastrophism* captures a problem that few have seriously grappled with. Anyone who wishes, as I do, for a new kind of (occupied) politics will have to face this formidable array of theoretically inspired reflections on the politics of apocalypse."
—Andrej Grubačić, author of *Don't Mourn, Balkanize!* and coauthor of *Wobblies and Zapatistas: Conversations on Anarchism, Marxism, and Radical History*

"Apocalypse and end-of-civilization memes are shaping and arguably undermining contemporary political and environmental movement organizing. *Catastrophism* unflinchingly challenges us, as movements, to seek alternatives to these narratives in determining our actions. It's a must read for anyone engaged in political organizing for truly long-term sustainable goals and futures."
—scott crow, cofounder of Common Ground Collective and author of *Black Flags and Windmills*

"This is a brilliant, timely book, a searching account of the limitations and inertia of catastrophic thinking. The authors urge us to move beyond doom-laden rhetoric in order to generate more ambitious analyses of our social crises. Above all, this book points the way toward fresh, energizing, and imaginative forms of social transformation."
—Rob Nixon, author of *Slow Violence and the Environmentalism of the Poor*

"In an age when even Mayan prophecies of the end of the long cycle are turned into prophecies of doom and destruction, this book offers a reasoned and lucid alternative understanding. Definitive and momentous, this book should be mandatory reading for everyone who wishes to comprehend the world we live in and change it for the better."
—George Katsiaficas, author of *Asia's Unknown Uprisings*

"Remember the story from *La Haine* about the guy who fell off a skyscraper? On his way down past each floor, he kept saying to reassure himself, 'So far so good . . . so far so good . . . so far so good. How you fall doesn't matter. It's how you *land*.' This collection of essays compiled in *Catastrophism* prepares us for that landing. Brilliant reading material for the abyss, and more: how to navigate an escape."
—Ramor Ryan, author of *Clandestines* and *Zapatista Spring*

"This important book aims to end the politics of The End. The authors of *Catastrophism* claim that apocalyptic politics, though promising to motivate revolutionary transformation, all too often leads to a fear-induced paralysis and cynicism. This book provides a badly needed boost to our political immunity systems against the apocalyptic claims bombarding us in this purported terminal year for our planet."
—George Caffentzis, author of *In Letters of Blood and Fire: Work, Machines, and Value in the Bad Infinity of Capitalism*

Editor: Sasha Lilley

Spectre is a series of penetrating and indispensable works of, and about, radical political economy. Spectre lays bare the dark underbelly of politics and economics, publishing outstanding and contrarian perspectives on the maelstrom of capital—and emancipatory alternatives—in crisis. The companion Spectre Classics imprint unearths essential works of radical history, political economy, theory and practice, to illuminate the present with brilliant, yet unjustly neglected, ideas from the past.

Spectre

Greg Albo, Sam Gindin, and Leo Panitch, *In and Out of Crisis: The Global Financial Meltdown and Left Alternatives*

David McNally, *Global Slump: The Economics and Politics of Crisis and Resistance*

Sasha Lilley, *Capital and Its Discontents: Conversations with Radical Thinkers in a Time of Tumult*

Sasha Lilley, David McNally, Eddie Yuen, and James Davis, *Catastrophism: The Apocalyptic Politics of Collapse and Rebirth*

Spectre Classics

E.P. Thompson, *William Morris: Romantic to Revolutionary*

Catastrophism: The Apocalyptic Politics of Collapse and Rebirth

Sasha Lilley, David McNally,
Eddie Yuen, and James Davis

Catastrophism: The Apocalyptic Politics of Collapse and Rebirth
Sasha Lilley, David McNally, Eddie Yuen, and James Davis

ISBN: 978-1-60486-589-9
Library of Congress Control Number: 2012913637

Cover by John Yates/Stealworks
Interior design by briandesign

10 9 8 7 6 5 4 3 2 1

PM Press
PO Box 23912
Oakland, CA 94623
www.pmpress.org

Printed in the USA on recycled paper, by the Employee Owners of Thomson-Shore in Dexter, Michigan. www.thomsonshore.com

Published in Canada by Between the Lines
401 Richmond St. W., Studio 277, Toronto, ON M5V 3A8, Canada
1-800-718-7201 www.btlbooks.com

Library and Archives Canada Cataloguing in Publication
Catastrophism; the apocalyptic politics of collapse and rebirth /
Sasha Lilley ... [et al.].
Co-published by PM Press.
Includes bibliographical references and index.
Issued also in electronic format.
ISBN 978-1-77113-030-1
1. Catastrophical, The. 2. Political psychology. 3. Expectation (Psychology)
I. Lilley, Sasha, 1970-

BD375.C37 2012 302'.17 C2012-903603-X

Published in the EU by The Merlin Press Ltd.
6 Crane Street Chambers, Crane Street, Pontypool NP4 6ND, Wales
www.merlinpress.co.uk
ISBN: 978-085036-632-7

Contents

FOREWORD

Dystopia Is for Losers

Doug Henwood

WHEN I STARTED WRITING THIS, NEW YORK CITY HAD JUST COME OFF its second punishing heat wave in three weeks. It broke with violent thunderstorms that prompted flash flood warnings from the Weather Service, spiced up with advice to those in low-lying areas to head to higher ground. Not two years ago, we had tornadoes that took down trees all over my neighborhood. Isn't this sort of thing supposed to happen in Kansas, not Brooklyn? Except that Kansas was in the midst of a huge, crop-destroying drought driving up food prices around the world.

The climate crisis has become part of daily life. It's no longer merely an abstraction of scientists' computer models—you can feel it when you walk out the door or when you shop for food.

But it's not only climate crisis that's becoming familiar. As I write this, the financial crisis that broke out in the summer of 2007 is about to celebrate yet another birthday. In the United States, the real economy began falling apart less than half a year later. We've officially been in recovery since mid-2009, but it hardly feels like it to most of us. This is clearly no mere cyclical affair, but a deep structural crisis of overindebtedness and profound maldistribution of income. Nor is it an American problem. Although the crisis broke out here, the epicenter has moved to Europe, whose neoliberal strategy centered on the euro project—in no small part an attempt to emulate the American model of looser regulation and "flexible" labor markets—is in collapse. In both New World and Old, the political system looks paralyzed in the face of the collapse, unable even to imagine a way out.

So, to paraphrase that remarkably banal phrase popularized by Alice Walker, these are the crises many of us have been waiting for. Weren't these the sort of crises that were supposed to wake up the somnolent masses and shock them into transformative political action? They haven't. What have they done? The maniacs that Richard Hofstadter wrote about in *The Paranoid Style in American Politics*—the ones who thought that Eisenhower was a socialist—now own a major party and a highly rated TV network. In Europe, social democratic parties impose austerity programs, only to be defeated by right-wing parties that do the same. Outside the mainstream, in the United States, the Occupy movement sprang up and then dissipated. In Europe, there have been numerous and inspiring demonstrations in Spain and Greece—though not in Ireland—yet the austerity programs, authored in Brussels and Frankfurt, roll on. On neither continent does anyone pay much attention to the unfolding climate catastrophe.

Catastrophe can be paralyzing, not mobilizing. Revolutionaries should be talking about possibilities of transformation, not spinning tales of great chaos and suffering. That's not to say that there isn't plenty of chaos and suffering in life. But looking to epochal quantities of both as the shocks that will awaken the masses out of their somnolence is not promising.

Hard economic times often don't help the good guys. In the United States, everyone thinks of the 1930s and the Flint factory occupations and gets weirdly nostalgic for the Depression. But the slump of the 1970s brought reaction. The "solution" to stagflation became Thatcher, Volcker, and Reagan. The right has done rather well in the current economic crisis too. In Europe, the 1930s—the decade of Hitler, Franco, and Mussolini—weren't at all kind to the left either.

On the contrary, good times are often better for the political strength of the masses. As Sasha Lilley says in her essay in this volume, "With the exception of the 1930s, periods of intense working class combativeness in the United States have tended to coincide with periods of economic expansion, not contraction and crisis. The two big strike waves of the early twentieth century, from 1898 to 1904 and 1916 to 1920, took place during years of growth." Some of the most intense political ferment in the world today—like factory and land occupations—is in China, which has been booming for decades.

The boss knows that there's nothing like a rise in unemployment to stifle militancy. This point was strongly made by the Polish Marxist economist Michal Kalecki in his great essay "Political Aspects of Full Employment":

> Under a regime of permanent full employment, the "sack" would cease to play its role as a disciplinary measure. The social position of the boss would be undermined, and the self-assurance and class-consciousness of the working class would grow. Strikes for wage increases and improvements in conditions of work would create political tension. . . . "Discipline in the factories" and "political stability" are more appreciated than profits by business leaders. Their class instinct tells them that lasting full employment is unsound from their point of view, and that unemployment is an integral part of the "normal" capitalist system.[1]

The bourgeoisie may understand the uses of crisis better than we do.

I can certainly understand the temptation of catastrophism. Faced with a population largely numb to environmental and economic disaster, one longs for some dramatic external intervention to do the work that conventional political agitation can't. So: the banking system will collapse utterly and people will finally wake up to the fact that the whole money and credit system is a sham and has been for at least a hundred years. Or we're going to run out of cheap oil, and the whole carbon-based energy system will collapse, and we'll all have to resort to growing food in our backyards (if you have a backyard, that is—and if you don't, that's probably your fault for living in an overpopulated, inauthentic city when you should really be scratching tubers out of the soil).

It's striking how much these doomy fantasies have in common with traditional right-wing thought. For decades, the right has denounced fiat money—a credit-based system organized around the state—as a crime against God and Nature, since the only "real" money is gold. As John Maynard Keynes put it, gold is "part of the apparatus of conservatism."[2] Curiously, a good bit of the left (you could find adherents around Zuccotti Park in late 2011, with their Ron Paul and End the Fed signs) has picked up on some of the analysis without embracing the rest of the apparatus—austerity and upper class power—that gold figures so centrally in.

The intellectual origin of population-decimating scenarios of ecological doom is that world-historical reactionary, Thomas Malthus. Engels described Malthus's vision as "the crudest, most barbarous theory that ever existed, a system of despair."[3] To that, many an environmental catastrophist would counter that Malthus had it right, even if he was a little ahead of his time.

But Engels was right. Catastrophism is a counsel of despair. James Davis has a lot more to say on that topic in his contribution to this volume, but I'll just note this: its native terrain is the right, which is all about natural limits (rather than social ones), great chains of being (with God at the top and the poor at the bottom, enjoying the suffering that is their lot in life), and punishments meted out for our sinful ways. Engels was writing from a position of revolutionary optimism that viewed scarcity and crisis as symptoms of a bad social organization that human intelligence and will could transform into something much better.

That sort of revolutionary optimism sounds quaint today. With the collapse of the USSR, we've not merely lost faith in transformative political projects, we often view them with fear: today's revolutionary will be in charge of tomorrow's firing squad, so let's junk ambition. Instead, it's as if many of us—and by "us" I mean those who long for a more peaceful, egalitarian society—project our transformative fantasies onto nature, but in a perversely destructive way. Nature will punish us for our ambition—"industrial society," the longing for material abundance, the urge to move beyond the small and local—by forcing us back to some hunter-gatherer purity, whether we like it or not. Short of that, there's always conventional financial collapse and pandemic economic ruin.

This sensibility is nicely captured by A.R. Ammons, in his book-length poem *Sphere:*

> man waited
>
> 75,000 years in a single cave (cold, hunger, inexplicable
> visitation of disease) only to rise to the bright, complex
>
> knowledge of his destruction![4]

Back to the cave!

So what can we set against this catastrophist trend? Or, as Eddie Yuen asks in his essay, "What narrative strategies are most likely to generate effective and radical social movements?" The question does make me a little nervous—it contains the temptation to follow George Lakoff down the road of "framing" and ignore the role of social power and inherited "common sense" in making some narratives more powerful than others. It's not just about coming up with a good story. Yes, Frank Luntz is a rhetorical genius for coming up with terms like "death tax," but there's also the centuries of money power in the United States that give the phrase more than rhetorical heft.

But narratives are important. Let's muse on one of the more famous exhortations in political history: "Workers of the world, unite. You have nothing to lose but your chains." It sounds quaint, almost embarrassing. It draws on structures of feeling that are looking pretty withered today: a notion of broad class solidarity and a sense that, once mobilized, this united class could transform the world into a better place. We're suspicious of solidarity because it erases "difference" and because we fear that it might universalize the perspective of the demographically dominant (e.g., educated white men). And we fear the transformation because of the firing squads (see above).

So instead, many radicals embrace small-scale efforts—the allegedly prefigurative little societies of the various Occupy encampments (which couldn't survive the police raids), or little co-ops, or any number of other tentative experiments in the interstices of the present. The question of scale is always elided.

How could these little things organize complex production and distribution for millions? How do we get those millions to play along with these experiments, given the distractions of daily life? And what precisely is the broad narrative appeal of DIY encampments when what people want to do is eat decently and be able to get their teeth fixed?

Not that it's easy to imagine mobilizing a large share of the population to an agenda that would mitigate the damage to the climate, much less reverse course in the future. Some on the left emphasize the responsibility of corporations for environmental ruin, which is true in part; there's no corner that a profit-seeking entity won't cut if it can get away with it. But getting serious about producing less carbon means that the cost of gasoline will have to rise, and Americans hate expensive gas. It would require different settlement patterns—less sprawl, less wasteful travel. It would mean living very differently. Strange weather might make people nervous, but they still have to drive to get to work if they've got a job that requires it. In a world with over a billion motor vehicles in use, how do you develop a constituency for driving less, or very differently? It's not easy to conjure up an answer. So it's tempting to think that catastrophe will force that constituency into being.

A variation of this approach in the economic realm, heard from many varieties of radical, is to claim that capitalism has no way out of this crisis. Catastrophe is upon us, and short of radical action, things can only get worse. Of course things could always get worse. It'd be foolish to say they couldn't. But capital typically finds its way out of crisis, otherwise the vast growth we've seen—global real income was up 13,637 percent between 1700 (just before the opening shots of the Industrial Revolution) and 2008—couldn't have happened.[5] Maybe it's different this time, and maybe capitalism has reached its ecological limits so no recovery will be possible. It seems an unwise bet, given the system's resilience and capacity to turn so many unimaginable things to profit centers. It could be a very ugly capitalism, reinforced with intensified state violence. Actually, given the intense repression that greeted mild Occupy protests, you could say that that's already happening.

But it doesn't seem fruitful to argue that there's no way to save the earth without ending capitalism. I would dearly love to end capitalism, but any strategy to reverse the despoilment of nature will have to happen before the system of private ownership can be transformed. It's possible that the development of those strategies—things like regulations, limits on the freedom to invest, the development and promotion of new energy sources—could help the work of transforming private ownership, but even under the best of circumstances it's hard to imagine getting the whole job done.

Mobilizing arguments about inevitable fates or cul-de-sacs without exit can demoralize more than they can rouse to action. If the climate can't be saved unless everything else is transformed, will that get people off their couches? Isn't the temptation, on hearing about an inevitable climate catastrophe, to mutter, "Why bother?" Why not just turn up the AC until the water's around your ankles? Or, to quote another bit from the Ammons poem I cited earlier:

> when one knows he's going out, can we
> blame him for shoving the voltage up?[6]

Wouldn't it be better to spin narratives of how humans are marvelously resourceful creatures who could do a lot better with the intellectual, social, and material resources we have? That new collectivities could together make a world better than the capitalist mess we've inherited? As someone who finds the temptation of pessimism too alluring, I keep reminding myself that recovering a utopian sensibility is about the most practical thing we could do right now. Dystopia is for losers.

INTRODUCTION

The Apocalyptic Politics of Collapse and Rebirth

Sasha Lilley

OURS IS AN EPOCH OF CATASTROPHE. NEAR-BIBLICAL FLOODS, HURRIcanes, and fires grow ever more ferocious and frequent, most lethally between the Tropics of Capricorn and Cancer. Financial havoc roils the North—likewise epic in nature, if not quite evoking Revelation—caught in the jaws of one of the most momentous crises of the capitalist system. An endless preoccupation with the end times, replete with the undead, weighs like a nightmare on the brains of the living. It would seem that only a correspondingly apocalyptic politics could measure up to the moment—driven by the urgent and warranted need, following Walter Benjamin, to sever the lit fuse before the spark ignites the dynamite—and prevent greater catastrophe. Yet in a world beset by calamity, might catastrophic politics end disastrously?

This book revolves around that most pressing question. In it, we explore the perils of what we term *catastrophism* within the environmental movement, the left, and the right, as well as examining the macabre visage of quotidian catastrophe.[1] Catastrophism presumes that society is headed for a collapse, whether economic, ecological, social, or spiritual. This collapse is frequently, but not always, regarded as a great cleansing, out of which a new society will be born. Catastrophists tend to believe that an ever-intensified rhetoric of disaster will awaken the masses from their long slumber—if the mechanical failure of the system does not make such struggles superfluous. On the left, catastrophism veers between the expectation that the worse things become, the better they will be for radical fortunes, and the prediction that capitalism will collapse under its own weight.

For parts of the right, worsening conditions are welcomed, with the hope they will trigger divine intervention or allow the settling of scores for any modicum of social advance over the last century. Not for nothing has the phoenix, rising muscularly from the ashes, been the far right's emblem. The environmental movement, by contrast, regards impending catastrophe with acute trepidation—and with good reason. But it tends not to grasp the crux of the ecological calamity and misses how fear can hinder, rather than help, its attempts to halt the disaster.

By its very nature, capitalism is catastrophic. There should be no doubt that the multiple social, and especially ecological, crises of our time are genuine and cataclysmic. We are suggesting, however, that politics embedded within the logic of catastrophe—that catastrophe will deliver a new world, or that it will create the conditions under which people automatically take action—do not serve the left and the environmental movement. An awareness of the scale or severity of catastrophe does not ineluctably steer one down the path of radical politics, in spite of received wisdom on the left and many great—albeit frequently dashed—expectations. Those who believe that the system will crumble from crises and disasters lose sight of the ways that capitalism uses crises for its own regeneration and expansion. Likewise, a focus on spectacular catastrophes typically overlooks the prosaic catastrophes of everyday life that are the sediment upon which capitalism is constructed.

The left has long held that a crisis can cause a rupture with the existing order, allowing people to throw off blinders and accreted prejudices that result from lifelong socialization. A cherished example is that of one of Pavlov's dogs, conditioned to salivate at the ring of a bell, who unlearned all training when its kennel flooded. But worse is not always better. Worse can just be worse. As partisans of the radical left, we are particularly concerned with how catastrophic politics can backfire on leftists and radical environmentalists. We also take a keen interest in catastrophism on the right, but for quite different reasons. While it short-circuits the left, right-wing catastrophism frequently helps shape the agenda of those in power.

Catastrophic politics within the left, right, and environmental movement are undoubtedly driven by disparate motivations. The left and right, after all, are not each other's *doppelgängers*, but antagonists, and the environmental movement straddles both. Yet fear binds them all together. Whether green, radical, or reactionary, catastrophists tend to believe fear will stir the populace to action. They emphasize panic and powerlessness, and conversely the vanguardist politics of the few. The politics of fear, however, play to the strengths of the right, not the left. Writers in this book argue that on the terrain of catastrophic fear, the left is not likely to win.

It should be said that critics of left- or right-wing catastrophism risk a certain smugness. It takes little effort to condescend to catastrophists as wild-eyed zealots, frantically hoping to upend the world, when simple reforms should presumably suffice.[2] Such is the typical reflex of liberalism, which regards catastrophism as the morbid symptom of left and right extremism, the politics of the fanatical outer edges. Yet liberals have their own brand of catastrophism.[3] The habitual reflex of liberal catastrophism is that if a right-wing candidate is not defeated in an election—usually by a center-right candidate—it will spell catastrophe. Hence, every four years, progressives in the United States put aside their misgivings with the choices on offer and canvass door to door to keep disaster at bay. Such fear-mongering in the service of the status quo reaches its apex with perennial liberal scares about impending fascism, with ceaseless invocations of the last days of Weimar Germany.[4] Such scares reached a fever pitch after the 2001 attacks on the World Trade Center and Pentagon, when various liberals decided that the Constitution had been ripped up, replaced by a dictatorship. More recently, disaffected liberal Chris Hedges, never a shrinking violet when it comes to summoning the apocalypse, has argued that there is a "yearning for fascism" in the United States. While opposing both major parties he warns that we may soon be "swept aside for an age of terror and blood," signaled by the violent rhetoric of Republicans.[5]

In *The Shock Doctrine*, Naomi Klein explores the political uses to which disasters have been put under neoliberalism. One could extend that timeline back to the beginning of capitalism. The state uses disasters and, furthermore, conjures up disasters—from war to fiscal crises—to drive through changes that otherwise would be untenable. Other radical writers have focused on the Kropotkinian moments of solidarity that can emerge after natural disasters.[6] In the wake of the Mexico City earthquake, Harry Cleaver wrote of the dialectic between the ways that those in power used catastrophe and the ways that the downtrodden organized themselves collectively in its aftermath.[7] In contrast, this book does not focus primarily on how catastrophes are used by the state and capital (although James Davis's chapter touches on this question as it relates to the catastrophic right). Instead we look at the role that catastrophe plays in the political rhetorics, and political choices, of the left, greens, and the extraparliamentary right. Our focus is on ideologies that are generated from outside of the state, even if they can be closely intertwined in the case of the right.

This book is a political intervention, designed to spur debate among radicals. We should be clear, however, what this volume is not. It is not an exhaustive or encyclopedic study of catastrophism or a set of instructions about what the left and environmental movement should do to politicize the apathetic and revitalize a broad anticapitalist project. While we point to the importance of mass radical organizing, it is beyond the scope of the book to explore where such movements have succeeded and failed in the past, and how that history relates to the appeal of catastrophic politics. We flag, instead, what we believe will not work—and what works at great cost. Catastrophic politics have a lengthy track record of failure. It is an approach destined for the blind alley.

The reach of this work is largely limited to North America and Europe. To be sure, the Global South has been at the receiving end of the most serious of catastrophes—from colonial plunder and empire to neoliberal structural adjustment, as well as the most severe effects of the ecological crises. It is our hope that others will explore catastrophism in the Global South.

★ ★ ★

Of all the alarming catastrophes at the beginning of the new millennium, the ecological catastrophe is without doubt the greatest and most serious: mass extinctions, ocean acidification, widespread and cataclysmic deforestation, the destruction of the coral reefs, number among many other horrors. Yet, as Eddie Yuen argues, the politics of catastrophe—expecting that knowledge of worsening catastrophe will arouse people to action—has foundered. In "The Politics of Failure Have Failed," Yuen draws on David Harvey's observation that mainstream Western conceptions of nature swing between cornucopian triumphalism about the power of science to master nature, and doom-laden pessimism over natural limits. Catastrophism inhabits the latter half of the binary, steeped in Malthusianism.

At the crux of environmental catastrophism, Yuen argues, are deeply held convictions about politicization. Most environmentalists operate under the assumption that if they are able to disseminate enough information about the dire state of the environment, the people will take action. The evidence suggests, however, that this assumption is wrong, and that increasingly urgent appeals about fixed ecological tipping points typically fall on deaf ears or result in greater apathy. Numerous apocalyptic scenarios that never came to pass have heightened the ineffectiveness of environmental rhetorics of catastrophe. One need only think of Paul Ehrlich's "population bomb" theories from the 1960s, in which horrible famine awaited the Global North, destroying England by the end of the twentieth century. Such false prophecies have inured some to the very real ecological crises of the present. A more recent example of false prophesy can be found with the millennium bug, or Y2K, scare, in which prominent greens like Helen Caldicott warned of impending nuclear meltdown and accidental war when the clocks struck midnight in the year 2000.

Most, but not all, ecological false prophecies originate in Malthusian premises, that absolute scarcity will cause various catastrophes. Yuen points out that their staying power has no correlation to their rate of predictability. Peak oil is the most prevalent form that this Malthusian current tends to take. Adherents posit that easily accessible petroleum reserves will soon peak—if they have not already—and that industrial society will unravel. Peak oilers wait with bated breath for TEOTWAWKI: the end of the world as we know it. Out of the collapse, a new sustainable, small-is-beautiful society may be born. As Yuen points out, such Malthusian notions about scarcity overlook the ways that capitalism has historically overcome obstacles and flourished from doing so. These crises fuel the system, opening up new rounds of capital accumulation.

Yuen notes, "The worst aspect of Malthusian scenarios however, is not that they are usually wrong, but that they tilt right. In fact, the predictable outcome of the Y2K and peak oil scenarios (were they accurate) is Hobbesian—'the war of each against all' and the legitimation of a militarized lifeboat ethics." He argues that unless environmentalists ground their politics in an awareness of class and geographical divisions, along with the ways that race and fear of those beyond one's frontiers are deployed, their catastrophic rhetoric may reinforce draconian environmental politics of elites who are less interested in stopping climate disasters than creating a lifeboat, or reinforcing borders, for the privileged. "Environmental catastrophism, unless it simultaneously argues that inequality, war, and imperialism compound the ecological crisis, is likely to encourage the most authoritarian solutions at the state level."

★ ★ ★

In "Great Chaos Under Heaven," Sasha Lilley traces the convoluted histories of catastrophic hopes on the left, looking at an interlinked, but seemingly contradictory, couplet of catastrophist ideas—one determinist, the other voluntarist. One half imagines that capitalism will hit absolute and insuperable limits and collapse under its own weight. The other presumes that worsening economic conditions or increased state repression will snap the somnambulant masses out of their slumber. Both sides have led to repeated predictions of impending massive crisis at every turn—sometimes borne out by reality and other times not. Both also have damaging political consequences, leading to quietism—the notion that there is little need for protracted organizing, or that deep-seated change will unfold mechanically—and adventurism—the idea that a small number of people can trigger revolutionary change. Not infrequently, the two are combined.

Marxism has been particularly bedeviled by the expectation of an automatic collapse of capitalism, dating back to influential debates within the Second International. Those who believed that capitalism was doomed trusted it would be brought down by inner imbalances, rather than class struggle. More recently, Immanuel Wallerstein has expounded the notion that capitalism is due for a collapse within the next twenty to fifty years. Although a collapsarian impulse runs through some of the left still—witness the triumphalism at the start of the so-called Great Recession that the system was crumbling—it has mainly fallen out of favor, done in by four decades of neoliberalism and the widespread view that there is no alternative to capitalism.

More prevalent now is the idea that worsening conditions are more auspicious for radicalization, whether through economic immiseration or the iron fist of the state. At its most notorious, it led to the rather overoptimistic notion that fascism's triumph in Germany would lead to communist revolution. While there is something to the trinity of crisis-war-revolution, the historical record shows that periods of crises, while polarizing, frequently spur people to move right, rather than left. And conversely, moments of relative affluence can, sometimes, enhance social power and the radicalism of demands. The point is that one cannot read the fates of social movements in the tealeaves of economic booms or busts. There is no automatic relationship to be found between the two.

The premise that worsening conditions are more fortuitous for radicalization has begotten a decidedly smaller strain of thought and action that concludes that if worse is better, then radicals should try to make things worse. The Weather Underground in the late 1960s followed this logic in its struggle against what it viewed as a fascist state lurking behind a liberal one. In our times, insurrectionism—both anarchist and antistate communist—attempts to heighten the contradictions of capitalist society through spectacular confrontations with the state. Some hope to trigger widespread insurrection, while others are content with doing the rebelling themselves.

Lastly, straddling both the determinist and voluntarist strains of left catastrophism, are the outlying ideas of radical opponents of civilization—some anarchist, some erstwhile anarchist. These radicals hope to hasten the inevitable collapse of industrial society, which they believe may be unraveling already. Civilization, for them, is catastrophic and only a catastrophe will bring it to an end.

What unifies these seemingly disparate political ideas and movements is an underlying despair at the possibility—and often the desirability—of mass radical or revolutionary movements. In some cases, it is driven by a will to power. The appeal of catastrophism tends to be greatest during periods of weakness, defeat, or organizational disarray of the radical left, when catastrophe is seen as the midwife of radical renewal. Such political despair is understandable. It needs to be resisted nonetheless.

★ ★ ★

James Davis highlights what he calls the disease-cure binary of catastrophe on the right in "At War with the Future." He points to the widely held conservative view that the gains of the left-wing social movements of the past have been catastrophic. Simply put, for the right social progress has been a disaster. In addition, distinct sectors of the right have seen an apocalyptic transformation, whether civil war or Armageddon, as the cure for the gains of the left. This binary of catastrophe as disease and cure structures his examination of the history of right-wing catastrophism.

Cure catastrophism, Davis suggests, takes on both religious and secular forms, although the ecclesiastical notion of the rapture is the most spectacular. He traces devout and profane forms, from dispensational premillennialism through to the imaginings of Norwegian mass-murderer Anders Breivik, who believed he was sparking a rebellion to rid the world of the damage caused by cultural Marxism.

Disease catastrophism, he argues, is a much broader category—perhaps universal on the right—that regards social progress as cataclysmic. He notes how a period of immense defeats for the left—the rise of neoliberalism, the crushing of organized labor, the rolling back of the gains of Civil Rights and the women's movement, the accelerated destruction of the environment—is registered by the right as nevertheless disastrous for reaction. The followers of the right believe they are losing. Right-wing catastrophists fret over cultural permissiveness, immigration, and in the United States and Europe, the presumed threat of Islam. Yet, as Davis argues, what far right catastrophists promote has been embraced by the mainstream; scares about immigration, Muslims, and borders are its standard fare. The tactics may differ, but the ends frequently do not.

Davis relates right-wing disease catastrophism to the actions of the state, which date back at least to Hobbes, and examines how such politics carve out space for the state to implement draconian policies—whether about foreigners, fiscal crises, or domestic terrorism. And, of course, the government can create its own states of emergency or use existing crises to its own benefit. "By framing questions like immigration as catastrophic problems, the state is able to respond with harsh and previously off-limits policies. Anti-immigrant sentiment is promoted throughout the European and American center right, and in both places border 'protection' and surveillance are expanding fiefdoms of the security state." As with environmental catastrophism, alarming rhetoric tends to bolster, not diminish, draconian state responses.

One might ask why right-wing catastrophism does not have the same paralyzing effects that Yuen flags in the environmental movement. Why do the foot soldiers of the right not burn out on false prophecy and the rhetoric of fear? Davis's answer is that catastrophism is not an ambivalent strategy for the right as it is for the left and among environmentalists. "From a rhetorical standpoint, catastrophism is a win/win for the right, as there is no accountability for false prophecy. On the one hand, it rallies the troops and creates a sense of urgency. On the other hand, though, fear and paranoia serve a rightist political predisposition more than a left or liberal one." Right-wing fear-mongering typically comes with scapegoats to blame, so the remedy is easily imaginable. Whereas the left frequently has a harder time identifying simple enemies and hence the route to a political solution of those fears is murkier. Davis notes that part of the success of the right is in filling the void vacated by those leftists who moved rightward and embraced neoliberalism. He posits that in a time of particularly pronounced economic insecurity, right-wing catastrophic politics can gain great traction.

* * *

Catastrophism allows us to lose sight, as David McNally argues, of the prosaic yet crucial catastrophes of everyday life that undergird the capitalist system. Capitalism creates spectacular catastrophes, but an emphasis on the spectacular alone—as with a focus on just moments of crises and rupture—only partially reveals how the system functions. McNally suggests in "Land of the Living Dead," the final chapter and a coda of sorts for the book, that it is precisely through anxieties and fascination with bodysnatching and the living dead that the quotidian workings of capitalism may be illuminated.

In recent years, the preoccupation with monsters, ghouls, and vampires seems to have reached a fever pitch in North America with incessant evocations of the impending zombie apocalypse. McNally connects these fears and fascinations of the past two hundred years with capitalism's relationship to the bodies of its workers projected onto the living dead, although the recurrent themes of bodysnatching and bloodsucking have become more overtly apocalyptic in our times. He suggests that this persistent fascination with the living dead is tied to the banal horrors of life under capitalism, which are overshadowed by the spectacular atrocities of the system but are no less fundamental to understanding it.

McNally examines the evolution of various incarnations of the living dead, originating in early capitalism. Mary Shelley wrote *Frankenstein* at a time when the bodies of the poor in England were being turned into commodities in both life and death—as wage labor for the living and, unwittingly, through a trade in body parts for the dead in a "corpse economy." The zombie took recognizable shape in the popular imagination under slavery on massive plantations in Haiti under French colonial rule—the living dead, toiling without consciousness. McNally traces the bifurcation of the zombie image into zombie-consumer and zombie-worker, the former a flesh-eater and the second an alienated laborer. As McNally shows, the first emerges again in sub-Saharan Africa under the dictates of neoliberalism, telling us a great deal about the system we inhabit. "For, in the picture of the maniacally insatiable flesh-eater, we find the capitalist-zombie, driven to relentlessly consume human beings. Meanwhile, in the image of the zombie-laborer we encounter the reality of the global collective worker reduced to a beast of burden who keeps the machinery of accumulation ticking. . . . Taken together, they define the zombie-economy of late capitalism, an out-of-control, flesh-eating machinery of manic accumulation and exploitation that has become an end in itself."

There is a positive utopian element, McNally suggests, embedded in the fleshy depths of the monster preoccupation. While the living dead conjure up the poor's fears of their literal or figurative dismemberment, they also evoke the fears that the rich have of the creature they have summoned up—the working class and poor rabble. McNally astutely observes that "The truly subversive image of zombie revolt in fact returns us to the everyday—to the idea that revolution grows out of ordinary, prosaic acts of organizing and resistance whose coalescence produces a mass upheaval. However extraordinary a popular uprising may be, it is nonetheless a product of decidedly mundane activities—strikes, demonstrations, meetings, speeches, leaflets, occupations." He goes on to add, "The apocalyptic scenario, in which a complete collapse of social organization ushers in a tumultuous upheaval, is ultimately a mystical rather than a political one. It is much more helpful to think about revolution as a dramatic convergence of real practices of rebellion and resistance that, in their intersection, acquire a qualitatively new form."

* * *

One might query the relationship between the different forms of catastrophism, especially the binaries of catastrophism on the left, the environmental movement, and the right. Is there a determinist and voluntarist couplet, similar to the one on the left, within the right? Undoubtedly right-wing catastrophism has its fatalists—such as those who believe that the Armageddon is inevitable—and its voluntarists—who believe they can bring on the collapse through concerted action. Likewise, does the fatalism of environmental catastrophism—one half of David Harvey's pair of Western triumphalism and pessimism—have a parallel on the left? As Raymond Williams noted, the left has its own binary of triumphalism and pessimism, although it would be an understatement to observe that the latter now predominates. However, an overarching logic might be harder to place, since these movements and ideas are driven by different impulses, from above and from below (and in the case of the greens, both). In that sense, it is hard to talk about a catch-all catastrophism, without specifying whether it is of the left, right, green—or liberal—variety.

Nevertheless, political despair and desperation do drive catastrophism. These are despairing times. For the left and for radical environmentalism, they are grimly urgent, despite remarkable acts of resistance—much of it collective—that have flourished of late. That despair is the culmination of decades of defeats and the retreat of a broadly utopian project—meant in the best sense of the word—that is committed to toppling capitalism through mass action. Catastrophism clings to the desire for a better world, while halfheartedly expecting to reach it through shortcuts. Like the zombie, it embodies despair and fear, as well as genuine and deep-seated longing.

Casting off despair, though difficult, is essential. We are now living through a crucial period of reorientation, when the radical movements are emerging once again. These are times to recognize defeats, without being paralyzed by them, and to assess what has—and has not—worked in the past. Catastrophic politics have a very poor track record. For radicals, it is high time to jettison them.

★ ★ ★

This work evolved out of many conversations within the Retort Collective, shepherded by Iain Boal. Retort is that most unusual of collectives, providing radical antinomian comradeship, but eschewing any party lines or set positions.[8] A number of us within Retort have independently set out from similar concerns about catastrophic politics, but arrived at quite far-flung destinations.[9] We would like to thank our fellow Retorters for their help in nurturing the project, without expecting any of them to agree with it. We also would like to collectively thank the meticulous staff at PM Press—Romy Ruukel, Brian Layng, Gregory Nipper, Joey Paxman, Craig O'Hara, Ramsey Kanaan, Stephanie Pasvankias, Jonathan Rowland, Dan Fedorenko, and John Yates.

Eddie Yuen would like to thank Azibuike Akaba, Iain Boal, James Brooke, Sean Burns, George Caffentzis, Chris Carlsson, Rosemary Collard, James Davis, Max Elbaum, Barbara Epstein, Silvia Federici, Betsy Hartmann, Ramsey Kanaan, George Katsiaficas, Joel Kovel, Sasha Lilley, David Martinez, Anne McClintock, Rob Nixon, Gene Ray, and all the great folks at Blue Mountain Center.

For having slogged through parts or all of the manuscript, and significantly improving it, Sasha Lilley would like to thank Terry Bisson, Iain Boal, Jim Brook, Jordan Camp, Jim Davis, Chris Dixon, Max Elbaum, Barbara Epstein, Andrej Grubačić, Karl Kersplebedeb, Gabriel Kuhn, Matthew Lyons, Donald Nicholson-Smith, Kay Trimberger, Cal Winslow, Eddie Yuen, and several anonymous readers. She is grateful to Summer Brenner, scott crow, Roxanne Dunbar-Ortiz, Laura Fantone, Juliana Fredman, John Gibler, Doug Henwood, Kathleen Lilley, Ted Lilley, Fouad Makki, Joseph Matthews, David McNally, Rick Prelinger, Charlotte Sáenz, Dan Siegel, Joni Spigler, Vanessa Tait, and Tom Wetzel for aiding the project in various ways. Most of all, she would like to thank Ramsey Kanaan for ensuring this book was not a total catastrophe.

James Davis would like to thank William Berkowitz, Tauno Bilsted, Fergal Finnegan, Jordan Camp, Sasha Lilley, Matthew Lyons, Eddie Yuen, Peter Linebaugh, Juliana Fredman, Kevin Coogan, Ramsey Kanaan, Cal Winslow, Peter Rudy, Whitney

Freedman, Iain Boal, Ramor Ryan, Shane O'Curry, Scott Fleming, Laura Fantone, Max Elbaum, DMZ, and Alan Toner.

David McNally would like to acknowledge Sue Ferguson—once again.

CHAPTER ONE

The Politics of Failure Have Failed: The Environmental Movement and Catastrophism

Eddie Yuen

THE SPECTRE OF APOCALYPSE HAUNTS THE WORLD TODAY. EVERY POLITical, cultural, and aesthetic field that we look at is replete with talk of catastrophe. This poses a particular challenge for environmentalists and scientists who are tasked with raising awareness about what is unquestionably a genuinely catastrophic moment in human and planetary history. Of all of the forms of catastrophic discourse on offer, the collapse of ecological systems is unique in that it is definitively verified by a consensus within the scientific community. The growing body of evidence is alarming. In addition to the well-known crisis of climate change, leading scientists have listed eight other planetary boundaries that must not be crossed if the earth is to remain habitable for humans and many other species.[1] These interrelated calamities include ocean acidification, the disruption of the nitrogen cycle, and the sixth mass extinction in planetary history, all of which are truly apocalyptic.[2] It is absolutely urgent to address this by effectively and rapidly changing the direction of human society. Unfortunately, discussion of this crisis and how to tackle it is often dominated by an undifferentiated catastrophist discourse that presumes apocalyptic warnings will lead to political action and hinders rather than helps the efforts of activists, scholars, scientists, and concerned people in general in bringing about the dramatic changes required.

In a world system saturated with instrumental, spurious, and sometimes maniacal versions of catastrophism—including right-wing racial paranoia, religious millenarianism,[3] liberal panics over fascism, leftist fetishization of capitalist collapse, capitalist invocation of the "shock doctrine," and pop culture cliché—what is the best way to articulate the all-too-real evidence for accelerating environmental catastrophe?[4] Is there, in fact, an inherently liberatory or radical politics that stems from a recognition of ecological catastrophe? If there is not, what effects do catastrophist rhetorics have on radical environmental movement building? As this essay will argue, even when dire environmental prognostications are accurate—and the evidence is overwhelmingly clear that they are—it is often the case that knowledge of "the facts" does *not* lead to an increase in political engagement. Given how high the stakes are, it is vitally important that environmental and climate movements understand the problems with catastrophism.

The foundational problematic of this book is the question of politicization: what narrative strategies are most likely to generate effective and radical social movements?

This essay will examine the main reasons that environmental catastrophism has not led to more dynamic social movements; these include catastrophe fatigue, the paralyzing effects of fear, the pairing of overwhelmingly bleak analysis with inadequate solutions, and a misunderstanding of the process of politicization. It will also explore capitalism's relationship to catastrophe and how the effects of environmental crises differ in their impact depending on place, race, gender, and class. The chapter examines how the long history of Malthusianism and previous false prophecies—doomsday predictions that did not come true—have shaped the current discourse. It explores the ways in which catastrophism may serve the interests of corporations. It concludes that unless some differentiation is made between antagonistic human communities, classes, and interests, environmental catastrophism may end up exacerbating the very problems to which it seeks to call attention.

We must start this inquiry by understanding that the veracity of apocalyptic claims about ecological collapse are separate from their effects on social, political, and economic life. One recent study found that, for many Americans, the more that is known about global warming, the less "personal responsibility" people feel for acting upon the crisis.[5] After surveying nearly 1,100 people, the authors state that "more informed respondents both feel less personally responsible for global warming, and also show less concern for global warming." They conclude that, "high levels of confidence in scientists among Americans led to a decreased sense of responsibility for global warming." Unfortunately, this evidence shows that once convinced of apocalyptic scenarios, many Americans become more apathetic. These studies illuminate basic political problems with the catastrophist rhetoric of the scientific and environmental communities. Why might their doomsday messages not be generating the desired results? This chapter is organized around several responses to this question.

Normalization of Catastrophe

> Western discourses regarding the relation to nature have frequently swung on a pendulum between cornucopian optimism and triumphalism on one pole and unrelieved pessimism not only of our powers to escape from the clutches of naturally imposed limits but even to be autonomous beings outside of nature-driven necessities at the other pole. . . . There is . . . nothing more ideologically powerful for capitalist interests to have at hand than unconstrained technological optimism and doctrines of progress ineluctably coupled to a doom-saying Malthusianism that can conveniently be blamed when, as inevitably they do, things go wrong.
>
> —*David Harvey*[6]

A common starting point for environmental catastrophism is that capitalist modernity is the best of all possible worlds, but is currently facing some *exceptional* problems. In this view, once these potentially disastrous problems are recognized, a combination of scientific innovation and popular belt-tightening should make possible a new period of growth without any fundamental changes.[7] Rather than seeing the various ecological crises converging now as exceptional, we must understand them as part of an *inherently* catastrophic mode of producing and reproducing social life. We must not take for granted the grinding, quotidian catastrophe of capitalism during the times when we are faced with exceptional calamities. This is especially true in our understanding of ecology, which has been profoundly shaped by the last five centuries of enclosure and commodification, a process that has accelerated in recent years.

Another pole of environmental catastrophism is that the current crisis is endemic to "civilization," or human nature itself. In some iterations, this also means that there is no differentiation between types of civilization, modes of production, culture, or technology. In some of these perspectives, all human activity is equally destructive, whether the mass extinctions caused to the "new lands" of Oceania and the Americas by Polynesians and Paleo-Indian or the current corporate ransacking of the planet by Chevron, Freeport-McMoRan, and RTZ.[8] This deeply pessimistic "primitivist" catastrophism places the problem too far upstream to speak meaningfully to the current crisis. The paradox of today's environmental crisis is that it is so tragically preventable: the great majority of capitalist production and consumption is patently unnecessary.

In the absence of a critique of the *specific* political and economic system in which the current ecological crisis is situated, the only solutions on offer will be moralistic and technocratic.[9] Worse still, there is a real danger that right-wing and nationalist solutions to the environmental crisis will become increasingly appealing. For these reasons, the stakes of accurately understanding the relationship between ecological and capitalist crisis could not be higher.

In her classic 1993 polemic against "apocalyptic environmentalism," geographer Cindi Katz argued that a politics of fear is rooted in the basic dichotomy of devastation or salvation, and ultimately breeds hopelessness. Overly generalized discussions of ecological collapse, for all their ostensible good intentions, tend to foreclose agency by functioning as a "totalizing narrative to end all totalizing narratives."[10] Historicizing the crisis does not diminish it. As Katz argues "contemporary problems are so serious that rendering them apocalyptic *obscures* their political ecology—their sources, their political, economic and social dimensions." Again, the issue is not the veracity of the science, but rather the larger politics within which the science is couched.

When we analyze the prevailing discourses on ecological collapse from an anticapitalist perspective, we better understand why many attempts at mass organizing have heretofore fallen flat. By pairing catastrophic information with glaringly inadequate solutions, the (majority of) scientific and environmental communities have offered little to inspire mobilization. Popular environmental films such as *An Inconvenient Truth* follow compelling evidence for ecological collapse with woefully inadequate injunctions to green consumption or lobbying of political representatives. The underlying message is that the only available form of political agency lies in being an individual consumer within the marketplace. For the same reason that a near plurality of Americans does not vote, ordinary people don't see "consuming virtuously" as a plausible solution. After all, why buy more expensive toilet paper or spend hours of unpaid labor separating trash when BP went back to making profits with oil drilling in the Gulf of Mexico not long after the Deep Water Horizon disaster? At best, such individualized response to the environmental crisis leads to existential, expressive, and voluntarist politics. A more common outcome, however, seems to be acute disempowerment and disengagement with environmental politics altogether. It is no wonder that the fear-based appeals to catastrophism favored by many environmentalists and scientists have not had the desired effects.

None of this critique is meant to disparage the remarkable work done by many environmental organizations, networks, and activists over the last few decades on issues ranging from anti-mining and anti-dam campaigns, conservation biology and biodiversity, protection of old growth and contiguous eco-systems, struggles to regulate and ultimately abolish toxic, nuclear and fossil fuel production, and many other issues. Were it not for this work, there would truly be no hope, and it is worth mentioning that environmental and climate justice perspectives are steadily gaining traction in internal environmental debates.

The Apocalypse Has Already Been Televised

It is a paradox of the twenty-first century that just as the contours of multipronged environmental crisis are coming into sharp focus, the world, and especially the United States, may be suffering from "catastrophe fatigue." Apocalyptic imagery has saturated popular culture for decades, but came to a boil with the "rapture" of 2011, the apocryphal "Mayan" prophecy of 2012, racist anxiety over the erosion of white majorities in the Global North, theocratic panic over the changing gender order, the ongoing financial meltdown, and the endless stream of "end-times" movies and video games.[11] The ubiquity of apocalypse in recent decades has led to a banalization of the concept—it is seen as normal, expected, in a sense comfortable. When a crisis does occur, people immediately reference it to movies, and there are now CGI images that serve as reference points for any conceivable disaster. Environmentalists and scientists must compete in this marketplace of catastrophe, and find themselves struggling to be heard above the din.

In this crowded field, increased awareness of environmental crisis will not likely translate into a more ecological lifestyle, let alone an activist orientation against the root causes of environmental degradation. In fact, right-wing and nationalist environmental politics have much more to gain from an embrace of catastrophism. This is especially true if the invocation of fear is the primary rhetorical device. Fear, as Rainer Werner Fassbinder pointed out, can "eat the soul." Fear is not a stable place to organize a radical politics, but it can be a very effective platform from which to launch a campaign of populist xenophobia or authoritarian technocracy under the sign of scarcity. Needless to say, fear is a logical and probably inevitable response to any person fully realizing the dire condition of the planet and its eco-systems right now. Emerging social movements will have to address this fear through a range of creative, directly democratic, and collective projects. This project is urgent, as environmental fears can be easily manipulated by capital and the state. Naomi Klein has famously described how the threat of economic disaster is a prerequisite for the "Shock Doctrine," and it is not hard to envision environmental correlates of this. An undifferentiated narrative of environmental doom is disempowering and encourages feelings of helplessness.

One useful model for comparison is the "scared straight" programs designed to steer teenagers away from drugs, gangs, and crime. Despite their fearsome reputation, these programs have been about as effective in intimidating working class youth from "high-risk" activity as abstinence only education has been in preventing teenage sex or DARE programs have been in curtailing drug use.[12] Such fear-based approaches fail in part because they are focused on changing individual behavior in the absence of structural critiques of the root causes of the problem (environmental crisis, addiction, crime, poverty, alienation, etc.). What good are moralistic and therapeutic proscriptions to social problems in the absence of more substantive, structural approaches to the dangers facing working class youth? By analogy, even if Americans were "scared straight" by Al Gore on the issue of climate change, what solutions does *An Inconvenient Truth* offer? The injunction to consume less or better (like the appeals to youth to refrain from sex, drugs, and gangs) is fundamentally at odds with the logic of post-Fordist capitalist culture that celebrates hedonistic accumulation unmoored from any "work ethic." For the earnest green consumer calculating his or her carbon footprint or the inner city youth wearing their chastity bracelets and "Just Say No" T-shirts, the prospects of "relapse" are quite high.

Why do most fear-mongering and doomsday scenarios have little to no politicizing effect at all? According to the aforementioned surveys, once convinced of catastrophic climate change, many Americans become *more* apathetic. To understand this, we must look to the conditions of atomization, depoliticization, powerlessness, and alienation that afflict the U.S. body politic generally. In these climate opinion studies, the only mentioned prescriptions center on individual consumption. To their credit, many people know better. They realize that individual consumer choice is largely irrelevant. For the same reason that people don't vote, Americans don't see "consuming virtuously" as a plausible solution.[13] This is a "glass half full" observation of sorts, as it shows that Americans can see through the façade of electoral politics, green consumerism, and blind faith in technocratic elites. It will remain a "glass half empty" situation, however, unless effective, participatory alternatives are realized. As it stands, undifferentiated environmental catastrophism leads to what Eric Swyngedouw calls a "post-politics" of administration by experts, and this will remain so until new forms of mass movements emerge. For many people, "waking up" in the context of alienation is profoundly disempowering, for the truth alone does not set one free.[14]

This outcome is but one aspect of a general confusion concerning the process of politicization in the last forty years. An unfortunate consequence of the extraordinary social eruptions of the 1960s was the template of politicization it established. The process by which millions of people "woke up" in the United States was in many respects unique to that era and is not replicable. This is due not only to the extraordinary wealth and rising expectations of that decade but, even more so, to the naïveté of that period which had itself emerged from the "clean slate" created by the repression of McCarthyism. For this reason, the shock of the JFK assassination, the shattering impact of *Silent Spring* by Rachel Carson, and the stunned disbelief greeting news of U.S. atrocities in Vietnam had a profound effect that cannot be replicated. Try as they might, activists cannot reproduce the revelatory disbelief of these traumas, the "loss of innocence," and the fervent activism that often followed. Many white Americans felt betrayed by a country that they felt had lied to them, but they still felt entitled and empowered as citizens to act.[15] The starting subjectivity of the last forty years could not be more different. A bitter skepticism toward government, media, and science is deeply rooted in publics worldwide, very prominently in the United States. In a culture of atomized cynicism, the doomsaying pronouncements of credentialed scientists will not cause the "scales to fall from the eyes" with regard to carbon emissions any more than the ghastly revelations of the Haditha massacre in Iraq in 2005 caused a revulsion comparable to the news of the My Lai atrocities of 1968.[16]

Periodizing *This* Catastrophe

> Di Eagle and di Bear a keep a living in fear
> Of the impending nuclear warfare
> But as a matter of fact, believe it or not
> Plenty people don't care whether it imminent or not
> Or who di first one to attack or if the human race
> [aba] survive or not
> For those whom is aware
> Them life already coming like a nightmare
>
> —*Linton Kwesi Johnson, "The Eagle and the Bear," 1981*

Perhaps the most optimistic assumption of environmental catastrophism is the notion that anyone who gets the message will work to prevent the onrushing disaster. Shocking though it may be, there is reason to believe that some business and political elites feel that they can avoid the worst consequences of the environmental crisis, and may even be able to benefit from it. The Hollywood film *2012* depicted a scenario of an "ark" of the planetary 1 percent escaping a global inundation, and this perception that there is a geographic or technical exodus strategy for elites is not an exaggeration. Although the greatest legacy of the George W. Bush administration may be the lost years of climate change denial, it's worth noting that Cheney and Rumsfeld had retreats high in the Rockies, where a business of high-end disaster survival shelters also thrives.[17] Although the insurance and reinsurance industries are extremely concerned about trillions of dollars in liability from predicted disasters, the opportunities for other sectors of capitalism are colossal in scope.[18]

Disaster capitalism, a term popularized by Naomi Klein to describe neoliberal shock therapy since the 1970s, has in fact been in effect from the very first conquests and enclosures of the sixteenth century. The expansion of capitalism has always been accompanied by differentially experienced social and environmental disaster—just as it has always been racial and patriarchal.[19] There is no evidence to think that present-day elites will see the suffering of the global majorities due to flooding and famine as any less "natural" than that of the millions who perished in the "late Victorian holocausts" documented by Mike Davis.[20] This is why catastrophist appeals to "humanity" from environmentalists and scientists haven't resulted in substantive action—many global elites have left "humanity" behind.[21]

Eric Hobsbawm has named the twentieth century the Age of Extremes, but the moniker "extreme" has become ubiquitous in popular culture only since the 1980s, used to describe everything from energy drinks, fashion, musical genres and sports to cuisine, nature shows and, of course, climate and environmental change. The pervasiveness of "extremity" is clearly a symptom of post-Fordism, "late" capitalism and, to use a very dated phrase, postmodernism. This should alert us to the ways in which the "extreme" times in which we live may not necessarily be the "end times." Put another way, the "end times" arrived for millions in the Western Hemisphere with the arrival of Columbus and countless species and eco-systems were condemned to make way for the "progress" enjoyed by the Global North. The sense of urgency now felt by many northerners is a very good thing, but in the long continuum of crisis, life for the Global majorities has long been "coming like a nightmare." The great Jamaican dub poet Linton Kwesi Johnson offers a still-useful distinction between the kinds of catastrophes feared in the Global North and South. People in the North were terrified of the bomb in the early 1960s and early '80s, but few were actually hurt by it. In contrast, during the same period in Guatemala, Ethiopia, Indonesia, Mozambique, Cambodia, and elsewhere, people lived with hunger, violence, and in some cases genocide—all threats far more immediate than nuclear war. A similar divergence in priorities is taking place now, and it is imperative that northern environmentalists understand this dynamic.

Whose Catastrophe?

> "At least the war against the environment is going well."
>
> —*bumper sticker seen in San Francisco*

> "The Planet is not dying; it is being murdered, and the people responsible have names and addresses."
>
> —*Utah Phillips*

In the late 1990s, the *Earth Island Journal* published an article titled: "Everybody Does Something about the Weather but Nobody Talks about It"—a clever inversion of an old phrase that reveals two things about climate politics. The first is that it remains a difficult subject to engage people, both because of its depressing nature but also due to the irreducible complexity and forebod-ing technical language. The second observation is that it is the default position of most scientists and environmentalists to speak of "everybody" as a unified subject. This seemingly innocuous "we" has had the pernicious effect of erasing the very meaning-ful class and geographic differences within humanity.

As noted earlier, much environmental discourse follows the logic of "scared straight": the notion that fear has a sobering effect. The assumption is that rational choices will follow a revelation. This assumption would perhaps be rational if humanity were undi-vided, and a representative global governance structure could be trusted to look after "the greatest good for the greatest number." Unfortunately, and contrary to the assumptions of many scien-tists, this is far from the case at the international as well as the national levels of decision-making. As it happens, the converg-ing crises facing humanity are occurring at a time when global wealth polarization is attaining levels not seen since the "gilded age" of Western imperialism. Four decades into the rollback of the Keynesian class deals of the Global North and the political sovereignty of the Global South, humanity faces a planetary crisis divided as never before. The richest 400 people in the United States have the wealth of the bottom 150 million.[22] Much of the litera-ture on climate change, peak oil, and other environmental crises do not hint at this division, however. Most scientists and journalists reflexively overlook class divisions in favor of telescoping their gaze to a higher level of abstraction, the species of Homo sapiens. This move, while refreshing when counterpoised to crude nationalism, assumes a potential unity that has yet to be realized.

To paraphrase the great Jamaican British critic Stuart Hall in a talk on race and racism: "I will assert my blackness until the clock strikes midnight, upon which time I will accept my universal humanity."[23] In this case, the premature assertion of "humanity" occludes clear-eyed scrutiny of the divisions that may put an end to all hope of that humanity. Beware of plutocrats speaking of Spaceship Earth.

Gleeful Doomsayers and Great Disappointments

Finally, environmental catastrophism has a long legacy of apocalyptic scenarios that were overwrought, exaggerated, or just plain false. False prophecies, like the proverbial boy crying wolf, lead to cynicism and burn out on the part of activists who may have become deeply invested in them. They may also discredit the authority of environmentalists and possibly science generally, if it is closely associated with a particular erroneous prediction. Such discrediting has often afflicted revolutionaries and religious prophets. One of the most amusing examples may be "The Great Disappointment" of 1844, in which followers of the Missouri preacher William Miller were gravely disappointed when the world did not end in 1844 and blamed him for their decision to abandon their families and property. In the same way that it is more illuminating to read last year's psychic predictions than the current ones, it behooves us to examine the litany of erroneous predictions made over the years, in order to put apocalyptic rhetoric in perspective.[24]

False environmental predictions fall into two categories. The first, and most ideologically significant, are Malthusian predictions predicated on the notion that absolute scarcity is the primary cause of ecological and social problems; we'll explore these in the next section. The second are predictions or concerns about potential environmental catastrophes that turn out to be false, for whatever reason.

Examples of the latter include the Russian physicist Valery Burdakov, who contended that the space shuttle would destroy the earth's protective ozone shield.[25] No false prediction has had greater effect on environmentally conscious people in the West, however, than the Y2K "crisis," the frenzy over the "millennium bug" that was potentially set to bring down the world's computer systems in the year 2000. Although initially postulated by computer scientists, the Y2K frenzy of 1999 took on a life of its own and brought together an unlikely assemblage of right-wing survivalists, white separatists, and goldbugs with anticivilizational luddites and primitivists, rusticating hippies, and antiauthoritarian leftists. Alternately sensationalized and ridiculed by the mainstream media, the scenario reinforced a deep-rooted skepticism toward technology and the state that transcends political lines in the United States. A surprising number of leftists and environmentalists bought into the prediction, and some gave money to the entrepreneurial right-wingers who had been stockpiling canned food, guns and ammo, bunkers, and other survival gear since the heyday of the John Birch Society.

Noble Prize–winning pediatrician and serial catastrophist Helen Caldicott took Y2K anxiety to its furthest extreme by arguing that the computer glitch was very likely to cause an "accidental Armageddon," including a possible accidental nuclear strike against the United States by Russia.[26] The following passage from a gold investment website encapsulates the feeling of those caught in the frenzy:

> If the polls are true, 15% of Americans will strip their bank accounts in December 1999. This equals a full blown banking panic. Probably about the same number will hit the grocery and hardware stores and strip them bare. This equals a full blown supply and distribution breakdown. The American government will then institute emergency measures that will then cause the other 85% of the population to panic. Terrorists will view the increasing chaos with the eye of an [*sic*] predator. A few selective strikes will cause a general feeling of hysteria in the land and across the globe-assuming we aren't already at

> war in the Mid-East or over Kosovo by then. This will equal a full blown social crisis.[27]

The long-forgotten Y2K saga is exemplary for revealing a crucial weakness in the way in which many environmentalists, liberals, and even leftists understand technology and the state. Technology is fetishized as both omnipotent and outside of existing social systems and modes of production. In hindsight, many people would have saved themselves time, money, political credibility, and anxiety had a better understanding of political economy been more widely circulated at the time. This would have made clear that capitalism accumulates precisely through overcoming technical obstacles. In fact, the "boom" due to hiring of all available technical personnel in the Y2K crisis provided a stimulus to the U.S. economy that was still reeling from the 1997 Asian and Russian financial crises. Then, as now, however, deep critiques of capitalism are not widely circulated in environmental, countercultural, and even progressive circles in the United States. Y2K is an interesting case study of the dead ends that catastrophism in the absence of such critiques can lead.

While making an inventory of recent false prophecies, it bears mentioning that there have also been "benevolent" predictions that have captured the imagination of some progressives and environmentalists. The Harmonic Convergence of August 1987 was a kind of New Age rapture that appealed to the same milieu of spiritual ecologists that are now attracted to Daniel Pinchbeck's apocalyptic reading of the Mayan calendar a quarter-century later. The story of the "Hundredth Monkey" gained significant purchase among peace activists in the 1980s and follows a logic that is a kind of mirror image of many catastrophic narratives.[28] According to this story, when enough monkeys in a group engage in a certain learned behavior (washing sweet potatoes), the behavior will jump to all related monkeys instantaneously, from mind to mind. The story was inspiring for peace activists, who saw in it the possibility that a change in consciousness could have a cascading effect. Although debunked by scientists, this and other "positive" prognostications were embraced by some environmentalists and share with Y2K and other "catastrophic" scenarios a kind of "substitutionism," in which a miraculous event (positive or negative) transforms consciousness, wipes the slate clean and abruptly changes the world with the need for difficult organizing and conflictive politics.

The Legacy of Malthus

Far more problematic than simply inaccurate predictions of catastrophe, however, are those that are steeped in theories of Parson Malthus—that is to say, in the common sense of capitalism itself. Malthus's theory that population growth will always outstrip the rate of growth of the production of food has long been discredited, but the fundamental assertion that overpopulation is the root cause of poverty and environmental degradation remains impervious to fact-based argument. As Eric Ross writes, Malthus's greatest achievement

> has been to provide an enduring argument for the prevention of social and economic change and to obscure, in both academic and popular thinking, the real roots of poverty, inequality and environmental deterioration. As such, no other ideological framework has so effectively legitimized Western interests, development theories and strategies, especially the Green Revolution and, now, genetic engineering in agriculture.[29]

None of this is to say that the sheer number of humans and domestic animals is not a huge environmental problem—it most certainly is. Overpopulation, though, is not the *primary* problem, nor is it the root cause of poverty, war, or human suffering. By shifting the responsibility for the crisis to the masses of poor people in the world, Malthusian environmentalists ease the pressure on the corporations, nation states, and militaries that are the true shot-callers in the world system. The status quo of capitalist production of unnecessary commodities and services for the global elites and "middle classes" is the ongoing catastrophe that must be addressed.

Strikingly, Malthusian catastrophism makes claims to be rigorously scientific and yet seems impervious to falsifiability, no matter how often its claims are disproven empirically. Critics of Malthus have long pointed to the convenient way in which Malthusian common sense "blames the victim" (especially women), and displaces responsibility for poverty, war, crime, and environmental destruction on the people who suffer most from these afflictions. But Malthusianism may also be comforting to wealthy and middle class interests precisely because its catastrophism forecloses any foundational critique of capitalism. Even if some environmentalists see the growth imperative of capitalism as "the logic of a cancer cell," they are comforted by the Malthusian common sense that poverty and suffering are natural and inevitable. Malthusian catastrophism justifies everyday triage of the global majority, and were it not for the fact-free impregnability of this doctrine, the global "middle class" would face an ethical crisis every day.

The simplistic Malthusianism behind much of the anxiety around "peak oil," resource scarcity, and overpopulation hinges on the everyday catastrophe of the poor (which is *natural*) metastasizing into a catastrophe for the well-educated denizens of the Global North as well. Scarcity becomes a crisis when those with "effective demand" are affected, and since this scarcity is "natural," the poor must be made to bear the burden of the "crunch" or simply go away. The impulse to blame the masses of the Global South for the environmental crisis is especially damaging, because they are often on the cutting edge of ecological activism and are looking for allies.

The worst aspect of Malthusian scenarios however, is not that they are usually wrong but that they "tilt right." In fact, the predictable outcome of the Y2K and peak oil scenarios (were they accurate) is a Hobbesian "war of each against all" and the legitimation of a militarized lifeboat ethics.

Apocalypse as Redemption?

> If the mediatised language of catastrophe is problematic it is because it is "apocalyptic" only in the Hollywood sense: it is devoid of ethical content. It says nothing of who we are and where we are going. This is something of a paradox: "apocalypse" is derived from the Greek word for revelation, or the unveiling of divine truth to mortal man. To many people apocalyptic literature (including biblical texts) represented the imaginative attempt to portray the corruption of the present in order to inspire radical social transformation. In contrast, populist catastrophism today represents a form of veiling, or clouding, of the ethical and political question of climate change. Perhaps what is needed, therefore, is more, not less, of the imaginative apocalyptic. This would frame climate change as an unfolding story in which we continue to play a part. And it would mean affirming the permanent ethical task of responding to the most despairing of situations.
>
> —*Stefan Skrimshire* [30]

In addition to assessing the impact of false predictions, it is also important to consider the impact of accurate ones, because this says a lot about the politics of catastrophism. Have environmentalists been given due credit when their apocalyptic warnings have been prescient, as with Rachel Carson's groundbreaking *Silent Spring* or the massive campaign against nuclear power prior to the Three Mile Island Disaster of 1978? Have the perspectives of working scientists been generally validated when they have accurately diagnosed and addressed a crisis, as with the curtailing of CFCs in response to the ozone hole in the Southern Hemisphere? What, if any, have been the political advantages of being able to say "I told you so"? The answer really depends on the deeper analysis offered by environmentalists and the overall political context. Certainly, environmentalists should claim credit if they are right with a catastrophic scenario, in order to buttress their credibility. For example, the destruction wrought by Hurricane Katrina, the recent fires and heat waves afflicting Europe, Australia, and North America, the BP oil spill, and the

Fukushima nuclear disaster were all forecast by environmentalists in broad terms.

It's unclear in these cases, though, if catastrophism is the most salient message. Unless a critique is made that links environmental crisis to social inequality and the "normal" functioning of the system, a steady drumbeat of catastrophism runs the risk of only pulsing from one crisis to the next.

This takes us back to our starting question: does catastrophism tend to be empowering or disabling for radical social movements? To answer that question, we need to examine how elites, the state, scientists, and activists on the right as well as the left deploy environmental catastrophism.

Catastrophe Capitalism

When environmentalists think of corporate influence on the climate debate, most immediately focus on the well-documented industry of climate change denial, best exemplified by the Koch brothers, Exxon Mobil, and various libertarian think tanks.[31] It is crucial to remember, though, that capital is divided on this question, and that far *more* corporate and elite energy has gone toward generating anxiety about global warming with an eye toward implementing "market-based" solutions. The late David Noble documented in 2007 how various elite organizations, led by Al Gore, the Pew Center, BP, DuPont, and other capitalist entities succeeded for a while in creating a "fevered popular preoccupation" with global warming.[32] The crucial difference between these elite campaigns and grassroots climate activism, however, was the corporate insistence that all inconvenient truths of capitalist exploitation and social inequality be swept aside. In the proposed "war" on climate change, rich and poor, North and South can be united through the purchasing power of our pocketbooks and our endorsement of an *increased* commodification and marketization of carbon and the atmosphere. In the Manichean contest "between mindless deniers . . . and enlightened global warming advocates," all critical scrutiny of corporate science must be suspended. In the "apocalyptic rush" to fight global warming, attention to empire, injustice, and free market theology are seen as distractions.[33]

In the wake of the Copenhagen climate summit debacle in 2009, the hot air has gone out of the carbon trading balloon and public concern over climate has plummeted to twenty-second place out of a possible twenty-two global issues.[34] Nevertheless, it's important to bear in mind that climate catastrophism has always had an elite component, and that apocalyptism was seen as the most effective way to generate popular support for a new round of enclosures and commodification. The project of Tony Blair, Bono, and Al Gore is most certainly *not* radical. Many if not most capitalists can recognize the overall environmental situation as catastrophic; the problem is that their proposed solutions exacerbate catastrophe for the global majority.

To better understand this, it is helpful to revisit some contributions of eco-Marxist theory. This goes beyond the common insight that capitalism has "the logic of a cancer cell" in that it must expand without end in order to exist, which, in a finite world, is unsustainable. Karl Marx made a more complex and elegant point that every boundary to accumulation that capital comes across is seen not as a limit, but as a starting point for devising a way to cross that barrier and initiate a new round of accumulation. As economist James O'Connor put it, "capitalism is not only crisis-ridden but also crisis-dependent."[35] In other words, capitalism produces terrible problems, but then escapes stagnation and enters its next round of accumulation by devising solutions to these self-same problems. At the level of the human body, capitalism has produced many new health problems (physical and mental), which impair the productivity of the workforce (cancers, nervous breakdowns, repetitive motion injuries, etc.). This barrier to accumulation (a degraded work force) is partially overcome by the proliferation of new health industries, pharmaceutical products, self-help books, etc. In a similar manner, ecological catastrophes encourage the proliferation of "green" industries (smoke stack scrubbers, biofuels, etc), nature based "derivatives" such as carbon trading, and the geoengineering industry, all of which are more likely to produce a new round of accumulation than actual solutions to the crisis.[36]

Capital, as its advocates argue, is indeed endlessly restless and innovative, but this restlessness is incredibly destructive as well.[37]

Catastrophe as Usual? Famine, Inundation, War, and the Triage of Humanity

Several dimensions of the climate crisis have geopolitical dimensions, and the rhetoric of elite and state environmental catastrophism reaches its apex in these areas. Among the projected catastrophes, one of the most frightening is the potential collapse of the food system. The most severe prognostications of climate induced food collapse project that four billion could die of hunger in the twenty-first century.[38] Due to the extreme political and social disorder that food insecurity on this scale would produce, militaries worldwide are already modeling scenarios based on such prognostications.[39] Among the countries most vulnerable to food crisis, the nuclear powers China and India are both facing catastrophic depletions of water and arable land. Needless to say, political and military elites are paying close attention to apocalyptic scenarios involving these countries.

In addition to the food crisis, the scientific consensus is that many coastal cities will become uninhabitable due to rising sea levels, including New York, London, Shanghai, and Miami. In particular, nine of the ten most endangered cities are in Asia; many of them are relatively new urban agglomerations that are home to tens of millions of people.[40] Long before actual evacuation is necessary, however, much coastal property will be uninsurable. This will have a massive impact on the world financial system, as the real estate values of many entire metropolitan areas will be massively written down. This scenario represents a convergence of environmental and economic catastrophisms, and is being vigorously discussed within the global reinsurance industry.[41]

These "natural" catastrophes immediately shift to geopolitical ones in the imaginations of pundits and planners in the rich nations, as the scenario of environmentally triggered violence and war is being vigorously "war-gamed" around the world.[42] Of any nation, Bangladesh is heralded as the maximum unavoidable climate calamity. Eminent paleontologist Peter Ward argues that all of the 150 million people currently residing in the country will be displaced by 2050, with the inevitable advent of an eight-inch sea level rise. Robert Kaplan, notorious for his paranoid post–Cold War polemic *The Coming Anarchy*, sees in Bangladesh the perfect melding of Malthus and Hobbes, with Islamic fundamentalism thrown in.[43]

The spectre of tens of millions of climate refugees clamoring across national borders by foot or boat is increasingly plausible, but most significantly it is also *familiar.* After all, millions of refugees from neoliberalism, neocolonialism, and war already circulate around the world. Rather than forcing elites to change their fundamental assumptions about the future of the current world system, such catastrophic scenarios encourage them to double down on the strategies of militarism and geopolitical realpolitik that are most comfortable. Needless to say, these strategies of nationalist bunkering, military "humanitarianism," and support for authoritarianism will have horrific consequences for the poor of the Global South. This is consistent, however, with the history of imperialism and globalization, even if this history is not so well known in the Global North. A triage of humanity has been taking place for five hundred years, and if the twenty-first century sees an increase in preventable death and misery, this will be more evidence of consistency than novel catastrophe.

This is the kind of catastrophe that indeed does have a silver lining for nation states, as a potential political unifier in the face of global calamity. The Keynesian stimulus hoped for by proponents of a Green New Deal is more likely to be a further expansion of border fences, naval patrols, military contractors, privatized security services, surveillance systems, and climate monitoring drones. To put it in the framework of Jared Diamond's bestselling book *Collapse*, there is no reason to believe that a civilization will change its ways when confronted with plausible prophecies of its demise. Environmental catastrophism (unless it simultaneously argues that inequality, war, and imperialism compound the ecological crisis) is likely to encourage the most authoritarian solutions at the state level. The chaotic world system that capitalism has brought into being ensures that no individual capital or state has the power or responsibility to counteract the tendencies toward ecological degradation. As Joel Kovel puts it: "Capital's regime of profitability is one of permanent instability and restlessness," and most political iterations of environmental catastrophism will only encourage a positive feedback loop of chaos and authoritarianism.[44]

A spikier, more paranoid version of elite climate catastrophism can be found in the various studies conducted by the U.S. military, its allies, and its think tanks in recent years. This approach, well documented by Christian Parenti in his book *Tropic of Chaos*, calls for military planning for the coming wave of resource wars and climate refugees that have already commenced in the Global South. The militarized solutions of the Pentagon seem downright rational, however, when compared to the outright denialism that can be found at high levels of the American political and corporate establishment. This denialism has set the stage for a shrill note of catastrophism sounded from the normally reserved and avuncular scientific establishment, led by figures such as NASA scientist James Hansen and Peter Ward.

Scientific Catastrophism

> The distinct millennialist discourse around the climate has co-produced a widespread consensus that the earth and many of its component parts are in an ecological bind that may short-circuit human and non-human life in the not too distant future if urgent and immediate action to retrofit nature to a more benign equilibrium is postponed for much longer.
>
> —*Eric Swyngedouw*[45]

The nightmare scenarios of Bangladesh in 2050, mentioned previously, are described by Peter Ward as a cautionary tale.[46] But what effect will this talk have on its audience? What does Ward enjoin us to do? He sees technical solutions as the last, best hope, and does not see fit to imagine a solution not at the level of the state. Ward, like James Lovelock, simply does not foreground the class and geographic differences in the world system. This is fundamental, since political and economic elites *do* see themselves as exempt from the crisis. To this extent, the aforementioned film *2012* is a surprisingly class-conscious take on apocalypse. Given the advance warning that world elites have of an impending catastrophe, elites literally produce a fleet of arks.

The key point is that, whether or not they are delusional, elites *think* that they are exempt and invulnerable.

The prevalence of fear-based catastrophism reveals the depth of acceptance of the assumption of rational choice theory in both the natural and social sciences. The assumption of a certain kind of instrumental rationality undergirds the delusional belief that if only people could understand the scientific facts, they will change their behavior and trust the experts. This essentially avoids an entire set of political questions that are the focus of this essay.[47]

In the United States, a collapsing educational system, deep seated anti-intellectualism, and a proud celebration of scientific ignorance typified by right-wing talk radio are commonly thought to be the wellsprings of popular resistance to scientific narratives. On the ethical level, many subjects of late capitalist consumer society can only extend compassion to individuals of a species or, at best, to domesticated animals or "charismatic megafauna" with which they are familiar. In a similar way, many people need a familiar "human interest" narrative to be able to relate to a "story"; this was typified by the massive public outcry over a U.S. soldier killing a puppy in Iraq, which contrasted to U.S. public indifference to the killing of tens of thousands of Iraqi civilians. Popular skepticism toward scientific claims is not always based on ignorance or alienation, however.[48] The long history of complicity between science and corporate and state power has created a deep climate of popular distrust that renders a simple revealing of environmental "truth" to be problematic. Now, more than ever, the need for a return to radical, democratic, and feminist science projects of the 1960s and '70s is vital if popular participation in environmental politics is going to be meaningful.

Austerity for All?

> *Con la rivoluzione caviale per tutti.*
>
> (After the revolution, caviar for everyone.)
>
> —*1970s Italian revolutionary graffiti*

What remedy do most scientific experts suggest? For many of them, austerity is seen as inevitable in any case and better embraced affirmatively than imposed through a "hard landing." James Lovelock, for example, calls for a "retreat" from industrial civilization itself, while some theorists and activists make a strong case for a managed downsizing of the scale of industrial civilization.[49] These proposals are appealing as the extraordinary stress on the world's eco-systems due to massive overexploitation is abundantly clear. The problem, as mentioned earlier, however, is that the decision-making and consequences of this program are highly unequal. The solution offered by global elites to the catastrophe is a further program of austerity, belt-tightening, and sacrifice, the brunt of which will be borne by the world's poor. Unsurprisingly, this program has not been embraced with enthusiasm by publics anywhere. In the Global South, two billion people do not have basic food security and it is obscene that the IMF, multinational corporations, and wealthy nations continue to demand even more immiseration and exploitation. Even for those not in abject poverty, the poor majorities do not see the justice of abstaining from their first experiences of automobilism, air conditioning, and consumer goods to order to pay the climate debt accrued by their former colonial overlords.

An effort toward a bottom-up and egalitarian transition is being embraced by ever-increasing numbers of people who are voluntarily engaging in intentional communities, sustainability projects, permaculture and urban farming, communing and militant resistance to consumerism. At the same time, however, politicians and corporate elites (the 1%, in the terminology of the Occupy movement) are aggressively imposing an increasingly undemocratic program of forced austerity aimed at ordinary workers, under the predictable mantra of TINA (there is no alternative). Under these conditions, class-consciousness in the Global North increasingly takes the form of resistance to austerity, and this presents a challenge to environmentalists. While it is certainly true that no amount of individual frugality will offset the carbon pouring into the atmosphere from industrial capitalism, it is also the case that the built in overconsumption of the Global North must be scaled back. This is why we must consider the alternative posed by the highly imaginative Italian left of the twentieth century. The explosively popular Slow Food movement was originally built on the premise that a good life can be had not through compulsive excess but through greater conviviality and a shared commonwealth.[50] Of course the Italian slogan "*Con la rivoluzione caviale per tutti*" cannot be taken literally: sturgeon populations are crashing worldwide due to the increased commodification of caviar for the trophyism of the banquet hall. But the spirit of joyful rebellion must be kept close at hand if the movements necessary for our era are to be fostered and encouraged.[51]

Catastrophism: Red, Black, and Green

As Antonio Gramsci famously put it, there are many morbid symptoms in the interregnum between the death of the old order and the birth of the new. In the current crisis, the "morbid symptom" of eco-nationalism is something to be constantly on the lookout for, as it threatens any compassionate or egalitarian approach to the crisis. This is because Malthusianism is at the core of most environmental discourse and Malthus's main political objective was to justify and naturalize the culling of the poor.

Like possessive individualism and sociobiology, these reactionary politics are at the present time thoroughly mainstream—they are in fact embedded in common sense. The *sine qua non* of right-wing catastrophism is the deployment of fear, and since there is nothing more frightening than climate change and its attendant calamities, the discursive terrain is ripe for reaction. Of course, political orientations never neatly fall into categories of left and right—in fact there are a range of often bizarre combinations, and there is a strong likelihood of many more "morbid symptoms" in the interregnum between the Holocene and the Anthropocene.

There are many alarming examples of racist, patriarchal, and xenophobic environmentalism, ranging from incoherent paranoia to outright eco-fascism.[52] It strikes many liberals as perplexing that bigotry and ecological awareness can coexist, but the history of environmentalism is rife with examples, ranging from the expulsion and historical erasure of Native Americans from National Parks, the Green wing of the Nazi Party, and the key role of eugenicist Madison Grant in the Save the Redwoods League.[53] While these kinds of politics may not currently have wide support, they can only be expected to increase once the disproportionate effects of climate chaos in the Global South begin to spill over to the North.

Catastrophism is rampant among self-identified environmentalists, and not without good reason—after all, the best evidence points to cascading environmental disaster. Warranted as it may be, though, the catastrophism espoused by many left-leaning greens remains Malthusian at its core, and is often shockingly deficient in its understanding of history, capitalism, and global inequality. Although it is assumed in U.S. politics that "greens" are socially liberal, the particular ways in which environmental consciousness is formed in the United States makes possible a particular insularity regarding class, inequality and, especially, race. As Carl Anthony has long pointed out, people of color in the United States have traumatic relations with land and nature in their collective experience. African Americans were bonded to the land for three hundred years,[54] Native Americans and Chicanos were violently expelled from their lands, and Asian Americans were "driven out" of rural areas to Chinatowns or internment camps.[55] American environmentalism has only recently been challenged by the environmental justice movement, and many self-identified greens are comfortable with a homogenous view of "community" and a single issue perspective on politics. All of this makes environmentalists extremely susceptible to simplistic analyses that place the blame for eco-catastrophe on an unspecified "human race" rather than particular agents in a particular social system. And if the human species is the problem and scarcity is the crisis, then the solutions to "eco-catastrophe" can look shockingly similar to right-wing prescriptions.

To give one example, James Howard Kunstler, in his book *The Long Emergency*, presents a neo-Malthusian scenario of peak oil and resource scarcity in which he advocates for closing the U.S. border with Mexico, a solution that would be welcomed by Lou Dobbs and Glenn Beck.[56] Kunstler's scenario, predicated as it is on a crisis of absolute scarcity as opposed to injustice and general instability, is indicative of the reactionary tilt of such arguments. The catastrophism that he espouses is driven by scarcity and cultural clash, not by neoliberal capitalism, and the only viable solution he sees is a bunkering of borders. Despite decades of anti-racist work by activists and academics, the fact that Huntingtonian "clash of civilizations" arguments can still have purchase in environmental circles is itself an alarming sign.

An even more pernicious form of green catastrophism can be found in some forms of primitivism. In the early twentieth century, anti-Semitism was said to be the socialism of fools. Does this make the advocacy of civilizational collapse the environmentalism of idiots? The idea that an abrupt end to industrial civilization, brought about by peak oil or some other Malthusian resource scarcity, could somehow encourage people to "live more lightly on the earth" is bizarre at best. Radical scholar Mike Davis once called Ernest Callenbach's *Ecotopia* the scariest book he'd ever read, given that it had no answer for what would happen to people of color in the event of a green secession of the Northwestern United States. Indeed, white separatists have for three decades espoused a "Northwest imperative" of a secessionist white republic in much of the same lands as *Ecotopia*. At the very least, those who see salvation in civilizational collapse should pay attention to the neo-Nazis who also harbor such fantasies. To quote political theorist John L. Schaar, when settling arguments among liberals, Hobbes wins every time. In other words, in a contest between armed survivalists and back to the land rusticators, the guns trump the organic chard.

Conclusion

"The thought of suicide is a great consolation," Nietzsche once wrote, "by means of it one gets through many a dark night." There is only one way in which the all too real environmental catastrophe can be seen as comforting, and this is for those greens who believe that it will bring about a transformative politicization of both ordinary people and elites. Yet, as this chapter has suggested, awareness or experience of environmental catastrophe doesn't necessarily lead to a progressive "awakening." The fear elicited by catastrophism disables the left but benefits the right and capital. In the case of environmental crisis, capital and the right's politics manifests itself as technocracy—on the one side of David Harvey's dichotomy of cornucopian optimism and unrelieved pessimism—and Malthusianism and eco-nationalism, on the other. Technocracy is the cornucopian side of the dichotomy; Malthusianism and eco-nationalism are on the other side of unrelieved pessimism. Both mainstream science and Western environmentalism are prone to elite catastrophism in these two variants—the former with a history of appeals to hierarchical techno-fixes and the latter with a long tradition of following a declensionist narrative (the fall from the Garden) that assumes an undifferentiated human race facing a common peril. It's an easy slippage from anxiety over losing what "we" have to an embrace of "lifeboat ethics," fortified climate enclaves, and a ceding of political power to technocrats. Shorn of a constitutive critique of the racialized and gendered dynamics of capitalist eco-destruction, the green capitalist and nationalist arguments could easily prevail.

Given the damaging and rightward-leaning effects of catastrophism, how would a critique of environmental catastrophism help reorient a radical environmental movement?[57] A dynamic environmental movement can't be based on critique alone, however. Naming and explaining capitalism as the root cause of the crisis is essential, but neither this nor an encyclopedic inventory of the environmental crisis will in and of itself call into being the solidarities, networks, and collective identities needed to bring about change. Any new movement must be rooted in networks of communities and activists who are engaged in self-organization—no organization or leader can conjure this into being. These new movements can't wait for capitalism to implode before offering solutions—solutions must be prefigurative and practical as well as visionary and participatory.[58]

A central lesson to take from the failure of catastrophism is that such a movement must make a positive appeal to community and solidarity, rather than a moralistic plea for austerity and discipline. As more and more people wise up to the false universality of calls for austerity, it is vital that a movement offer something positive to go with the cold porridge of climate catastrophe. This something, for inhabitants of the Global North, is an opportunity to escape alienation and exploitation for a chance to build something new. The ecological catastrophe will not automatically lead to a better world. But in organizing against climate apartheid, and the enclosure and commodification of nature, one hopes that a compassionate, egalitarian, and radical movement can help bring that new world into being.

CHAPTER TWO

Great Chaos Under Heaven: Catastrophism and the Left

Sasha Lilley

WHEN THE FINANCIAL CRISIS UNFOLDED IN THE EARLY YEARS OF THE new millennium, many radicals hoped the end was nigh. The capitalist system had finally arrived at its terminus—the last stop on the line. Where the left had failed, the inexorable limits to capitalism would deliver. Needless to say, it has not turned out so well.

There is nothing eternal about capitalism. It came into being fairly recently, and in all likelihood will be ushered out one day. Yet the belief that it will come crashing down without protracted mass struggle is wishful thinking. Expecting predestined forces to transform society for the better forms one half of the couplet of left catastrophism, which has shaped the radical tradition for well over a century.[1] The other consists of the idea that the worse things get, the more auspicious they become for radical prospects—that if economic conditions or political repression become dire, the scales with fall from the eyes of the misled masses, allowing them to put an end to a much greater catastrophe. This notion is beset with similar defects. Periods of radical social upheaval have followed economic crises and—especially—war. But there is nothing preordained about this relationship.

The magnetic pull of catastrophic ideas is indisputable—often, but not always, when the forces of the status quo seem firmly entrenched and the forces of the left are in disarray. In a time of pronounced—or at least more visible—crises, their attraction appears stronger. This chapter traces the contours and consequences of the left catastrophist dyad—not from a vantage point of liberalism, or an animus to revolutionary change, but rather from one that shares the aims of radicals seeking an end to the capitalist system. While the deterministic version of catastrophism holds less sway now, after decades of disappointed predictions of the terminal crisis of the capitalist system, the voluntarist variant flourishes and takes a number of forms, all predicated on the idea that the worse things get, the more favorable they will be for the left. In the furthest reaching version of this logic, left-wing catastrophism beckons—or wills—the collapse or destruction of industrial society in toto. (Hence, the essay begins with Mitteleuropean debates about economic breakdown and comes to a fiery end with socialism via nuclear annihilation.)

Rather than functioning in stark opposition, determinism and voluntarism can overlap, their adherents oscillating between the two in a dialectic of disaster.[2] In this essay, I argue that, irrespective of the form, what binds together this catastrophic dyad of determinism and voluntarism is a deep-seated pessimism about mass collective action and radical social transformation. Political despair lies at the center of left-wing catastrophism. The defeats underlying such despair are complex and beyond the scope of this essay. Neither does it attempt to furnish prescriptions for mass action and revolt. What it does point to is what does not, and will not, work. A militant radicalism with any prospects of success embraces catastrophism at its peril.

This essay focuses on catastrophism within the two major traditions of the radical left in the Global North, Marxism and anarchism, but in no way comprehensively surveys their entanglement with catastrophic politics.[3] Some may object to whom the label "left" is affixed here, making unwitting bedfellows of Stalinist parties and libertarian communists, Maoists and anarchists; some, it could be argued, are the gravediggers of revolution, not its champions. Yet all either describe themselves as members of the left or come out of that tradition. Whether we like it or not, they have shaped its legacy.

I focus on the left in the Global North, particularly in the United States and Europe, for reasons of proximity and familiarity. Catastrophism in the Global South is exceptionally complex, generated to some degree by similar ideological and political impulses as the North, yet also by a sharply different context.

Catastrophes in the Global South, where most of the world's inhabitants live, have occurred at much larger orders of magnitude than almost anything experienced in the North. One does not need to go back to the genocide of native peoples in the Americas, late Victorian holocausts, King Leopold's slaughter of the Congolese, or even the *Nakba*—or "catastrophe"—in which two thirds of the Palestinian people were expelled from their homes. In the last three decades, over one hundred thousand people were killed in the U.S.-backed war in Guatemala; a third of the population of East Timor was slaughtered during its battle for independence from Indonesia; hundreds of thousands were killed in the violence unleashed by the U.S. invasion of Iraq; and more than 5.4 million people have died in a war that spanned much of Central Africa—more deaths than any other conflict since World War II.[4] In the Global South, protest against catastrophe has taken a range of forms, some based on hope and strategic resistance, others stemming from despair. Like the late nineteenth-century ghost dance of the Plains Indians—turning inward in their desperation, their populations wiped out—the peoples of the Global South have often been swallowed up by the politics of despair. The U'wa people of Colombia and Venezuela, who threatened to commit mass suicide in 1997 if Occidental Petroleum drilled on their land, may represent the most dramatic example of this. Such politics need extensive and nuanced exploration, which is unfortunately beyond the scope of this work.

Breakdown Theories of Capitalism

"Irresistible economic forces lead with the certainty of doom to the shipwreck of capitalist production," mused Karl Kautsky, doyen of the socialist Second International, in 1892. "The substitution of a new social order for the existing one is no longer simply desirable, it has become inevitable." Kautsky's ebullience arose from his early certainty that capitalism would be rent asunder by insurmountable constraints.[5] The notion that capitalism would inevitably collapse under its own weight makes up the determinist half of the catastrophist dyad, premised on the presumption that capitalism will butt up against insuperable economic limits.

The idea that capitalism would come to an end is not unique to Marxism, and was particularly prevalent in the first half of the twentieth century; economist Joseph Schumpeter—briefly Austria's minister of finance—concluded that capitalism would eventually render itself obsolete.[6] Yet the Marxist tradition—at least prominent parts—has been particularly afflicted with the notion that capitalism would catastrophically break down rather than be vanquished by the downtrodden.

Marx is often remembered as a prophet of capitalism's demise, brought down by the iron laws of history. Crises were to lead to ever greater crises and finally to the expropriation of the expropriators. In the midst of the 1857–1858 crash, Marx and Engels were swept up in pre-Kautskyesque jubilation, while Marx worked on *The Grundrisse.* "Though my own *financial distress* may be dire indeed, never, since 1849, have I felt *so cozy* as during this *outbreak*," Marx wrote his collaborator. Engels replied two days later: "In [the revolutionary year] 1848 we were saying: Now our time is coming, and so *in a certain sense* it was, but this time it is coming properly; now it is do or die."[7] Yet, despite Marx's early enthusiasms and the declarations of his critics and supporters, his crisis theory did not pivot on capitalist breakdown.[8] While Marx argued that crises are essential to capitalism, he did not equate such crises with the collapse of the system. His writings in *Capital* on the tendency of the rate of profit to fall did not presume an inexorable path to a final collapse of the system, but rather a continuous back and forth between such a tendency and countervailing forces.[9] Rhetorical flourishes aside, Marx believed that class struggle and collective action, not the unfolding of predetermined laws, would bring an end to capitalism. "*History* does *nothing*," Marx and Engels emphasized in *The Holy Family*, "it 'possesses *no* immense wealth,' it 'wages *no* battles.' It is *man*, real, living man who does all that, who possesses and fights; 'history' is not, as it were, a person apart, using man as a means to achieve *its own* aims; history is *nothing but* the activity of man pursuing his aims."[10]

Despite—or perhaps owing to—the complexity of Marx's understanding of crisis and the impermanence of capitalism, many of his successors have interpreted his writings only selectively. At the turn of the last century, they deemphasized his concept of class struggle in favor of a mechanistic conception of the system's disintegration on one hand, and reformist evolution on the other.[11] Debates that roiled the Second International during a period of economic expansion preceding World War I became the touchstone for future notions of capitalist collapse. The lynchpin of the International, and Europe's largest socialist organization, the German Social Democratic Party (SPD), was convulsed over the possibility of an evolutionary road to socialism. Party reformists, led by erstwhile bank clerk Eduard Bernstein, held that crises were not endemic to capitalism, and surmised that the system could blossom into socialism through parliamentary means. Bernstein's radical adversaries contended that the orthodoxy of the SPD—that capitalism was not simply crisis-prone, but that it would buckle from internal contradictions—was correct.[12] Proponents of *Zusammenbruchstheorie*, or breakdown theory, were justly alarmed by the rightward turn in the SPD. Nevertheless, they believed capitalism's fate would be inevitably sealed by impassable limits, rather than through the flesh-and-blood strife of class conflict. Their particular reading of Marx has colored the communist tradition to this day.[13]

Even a thinker as brilliant and nuanced as Rosa Luxemburg maintained that capitalism would abut and then collapse against absolute limits. In *The Accumulation of Capital* (1913), she posited that capitalism could only sustain itself via imperial expansion, by finding noncapitalist markets in the territories brought to heel by colonial conquest. Without new noncapitalist markets, capitalism could not survive. Once imperialism conquered the globe, capitalism would face a mortal crisis, conducting the catastrophe of empire back to the omphalos of the capitalist system.[14] "Imperialism brings catastrophe as a mode of existence back from the periphery of capitalist development to its point of departure," she argued. "The expansion of capital, which for four centuries had given the existence and civilization of all noncapitalist peoples in Asia, Africa, America, and Australia over to ceaseless convulsions and general and complete decline, is now plunging the civilized peoples of Europe itself into a series of catastrophes whose final result can only be the decline of civilization or the transition to the socialist mode of production."

Unlike her contemporaries within the SPD, Luxemburg's confidence in the inevitability of a collapse did not lead her to political quietism, or inaction. On the contrary, she argued that workers in the imperial countries must oppose empire and organize for socialist revolution before the inevitable breakdown results in barbarism.[15] In doing so, Luxemburg inverted the typical trajectory of catastrophic determinism: instead of capitalist collapse heralding a new society, it will produce barbarism unless revolutionaries achieve socialism first.

Following the Second International's split; revolutions in Russia, Germany, Hungary, and Italy; and the ascendancy of fascism in the latter, debates were recast within the Communist movement about whether capitalism would meet an internally generated demise. While Lenin dissented, the notion became firmly rooted in the orthodoxy of the increasingly Stalinized Third International, with dire political consequences.[16] It remained a constant thread despite the twists and turns, and changes and reversals, of the Comintern's strategic outlook.[17] In the age of imperialism, it was surmised, capitalism was no longer capable of developing the forces of production. Unable to renew itself and rotten within, it would ultimately break apart. "The general picture of the decay of the capitalist economy is not mitigated by those unavoidable fluctuations of the business cycle which are characteristic of the capitalist system both in its ascendancy and in its decline," declared the "Theses on Comintern Tactics" passed by the Fourth Congress of the Comintern in 1922. "What capitalism is passing through today is nothing but its death agonies. The collapse of capitalism is inevitable."[18]

Various political lessons were drawn from this, albeit haphazardly applied, depending on which way the winds of Stalin's fancy were blowing at the time. If capitalism was in its death throes, socialist revolution must necessarily follow as the next stage in history. Hence, except for periods deemed temporary moments of stabilization, revolution was in the cards—not because of the intensity of class struggle, or the extent of mass organization, but because capitalism had become overripe. R. Palme Dutt, the Anglo-Indian theoretician of the Communist Party of Great Britain, wrote in 1931:

> On a world scale, each crisis gives place only to a greater crisis. There is no way out of this succession within capitalism. The hold on the Empire, on one quarter of the world, by British capitalism helped in the past to hide the decline for a while, to give new sources of profit and tribute. But to-day this hold is breaking down. The Dominions move away on their own capitalist development. The Colonies revolt. India is fighting its way to freedom. Once this hold is lost, what is the future for British Capitalism? How shall the forty-five millions in these islands live? There is no future on the basis of capitalism in Britain. Unless we overthrow capitalism, not only decline, but final collapse and starvation await us.[19]

The mechanical imminence of revolution led the Comintern to support insurrections during politically inauspicious moments with little planning or mass organized support.[20] "The most tragic expression of this exaggeration [that final victory was approaching] came in Estonia," wrote C.L.R. James, "where at 5.15 am on December 1, 1924, two hundred and twenty seven Communists started a revolution, and by 9 o'clock were completely defeated, doing untold harm to their own party and the idea of proletarian revolution around the world."[21]

More than a decade after the apex of the breakdown debate—and mere months before the 1929 stock market crash—the Polish Marxist Henryk Grossman published *The Law of Accumulation and Breakdown of the Capitalist System*. Grossman, who was affiliated with the Frankfurt School, reignited the controversy by arguing that Luxemburg was correct in stating that the accumulation of capital runs up against absolute limits ending in breakdown, but wrong in her reasoning.[22] He contended that the limits from which capital would collapse were internal to the sphere of production, rather than circulation: "Despite the periodic interruptions that repeatedly defuse the tendency toward breakdown, the mechanism as a whole tends relentlessly toward its final end with the general process of accumulation. . . . Once these countertendencies are themselves defused or simply cease to operate, the breakdown tendency gains the upper hand and asserts itself in the absolute form as the final crisis."[23]

Yet even in Grossman's time, at least one radical voice challenged his collapsism. Council Communist Anton Pannekoek rebutted Grossman's analysis, which had become influential among left communists in the 1930s.[24] He charged Grossman with citing Marx's writings on capitalism's periodic crises to suggest the latter believed in the notion of a collapse from its limits. Pannekoek argued that the system needed to be overthrown because of capitalism's durability, not its weakness: "It is not due to the economic collapse of capitalism but to the enormous development of its strength, to its expansion over all the Earth, to its exacerbation of political oppositions, to the violent reinforcement of its inner strength, that the proletariat must take mass action, summoning up the strength of the whole class." "The self-emancipation of the proletariat," Pannekoek concluded, "is the collapse of capitalism."[25]

Pannekoek reasoned that Grossman's theory appealed to left communists partially because the expectation of collapse handily foreclosed the debate over whether capitalism could be reformed.[26] If the system were in its death throes, reformism would become redundant. But he concluded that radicals see the masses as "passive and immobile," caught in old politics, and that immobility fuels the appeal of theories of collapse "The few revolutionaries who understand the new development might well wish on the stupefied masses a good economic catastrophe so that they finally come out of the slumber and enter into action." Yet, Pannekoek observed, "struggle is never so simple or convenient, not even the theoretical struggle for reasons and proofs."[27]

In our times, the idea that capitalism will run up against impassable limits is perhaps best represented by the neo-Marxist sociologist Immanuel Wallerstein. He has long prognosticated the demise of the capitalist world system along with the decline of U.S. hegemony. Its death date, however, has been moved forward a number of times. Wallerstein posits that the capitalist world system, unable to reap sufficient profits, has been stagnating since at least the 1970s. Such stagnation, he argues, follows larger patterns of economic expansion and contraction dictated by Kondratiev waves.[28] Wallerstein maintains that rising production costs relative to prices—primarily the diminution of low-wage areas globally, an increase in taxes, and the increasing difficulty in externalizing production costs by dumping toxic byproducts onto the environment—are eroding the profits of capitalists.[29] Accordingly, a new world system will emerge to replace capitalism as it runs up against such fixed limits. Twenty to thirty years from now the capitalist world system will no longer be with us: "We can have a system better than capitalism or we can have a system that is worse than capitalism. The only thing we can't have is a capitalist system."[30]

Given capitalism's history of overcoming obstacles to accumulation, however, it is hard to see why rising production costs would be absolutely insurmountable. And if capitalism were not able to find ways to cope with these problems, it is not clear why the system would collapse as a result.

Imagining that there are no constraints to political action in any given moment—that one can conjure up revolution on the strength of one's determination and outrage—is fraught with problems, as we shall see. But expecting limits to accumulation, the unfurling of iron laws of history, or the descending trajectory of economic waves to cause capitalism's demise is equally unwise.[31] The cardinal strength of capitalism is its immense and terrible dynamism, its ability to circumvent roadblocks and constraints. Rather than coming to a halt in inescapable cul-de-sacs, capitalism finds ways out by opening entirely new avenues to accumulation, frequently by generating new needs and desires. The crises for which many collapsarians eagerly await, expecting the whole edifice to keel over, are in fact moments of destruction in which the system renews itself and sets the stage for the next crisis.

Radical expectations of collapse frequently lead to the twin dangers of adventurism (the ill-conceived actions of the few) and political quietism (the inaction that flows from awaiting the inexorable laws of history to put an end to capitalism). In both, class struggle is sidelined. The prognosis that the system will keel over originates—at least theoretically—in an impoverished and mechanical understanding of capitalism. Antonio Gramsci, whose *Prison Notebooks* ruminated on the experience of revolutionary defeat in Italy, emphatically decried mechanistic politics. Gramsci believed that "while capitalism might enter economic crises they are not life threatening. At best they offer a more favorable terrain for the dissemination of socialist ideologies but were most likely to be the vehicle through which capitalism restructured itself."[32] The work of anticapitalist political organizing, as long and arduous as it is, cannot be replaced by the shortcut of an irrevocable system failure or by plucking revolution out of thin air.

It bears mentioning that the idea of a mechanical breakdown of capitalism has frequently been shadowed by the notion that capitalism, if not on the verge of collapse, is moribund. In the 1880s, in the midst of a protracted economic crisis that had not managed to bring down the system, Engels concluded that capitalism was stagnating. This handy diagnosis would capture the imaginations of Marxists during two periods of economic prosperity and capitalist expansion: in the decades preceding World War I and in the boom following World War II. Lenin and Zinoviev adopted much of this argument from liberal economist John Hobson's *Imperialism*, published in 1902. The latter maintained that imperialism was driven by a search for new markets, owing to insufficient demand for goods within the imperial powers caused by wealth inequality.[33] The notion that capitalism was stagnant was revived during the prosperity of the post–World War II era, when capitalism had not only survived but also appeared to be expanding. Economist Paul Baran, associated in his early years with the Frankfurt School, and his collaborator *Monthly Review* founder Paul Sweezy, believed that capitalism in its monopoly form was mired in chronic stagnation, based on the inability of capitalists to find profitable avenues for investment and the inability of workers to buy.[34] They believed that state intervention—especially military spending—could postpone the mechanical breakdown of capitalism in its monopoly form, but that collapse would one day be inevitable. Baran and Sweezy did not see the working classes of the advanced capitalist countries as capable of bringing capitalism down, but—like many radicals to follow them in the New Left—looked instead to the Third World.

Such a stagnationist perspective, which continues to have wide appeal today within the (not very wide) world of Marxist economists, appears to be beset by some of the same problems as the notion of collapse. In particular, they both revolve around the idea that capitalism is chronically infirm. Yet capitalism seems to survive and expand (and periodically and destructively contract), whether or not critics decide it is stagnating. These notions seem colored by the assumption that the wider public will not oppose capitalism unless it can be convinced that the system is on the way out. Perhaps we need to recognize that a radical anticapitalist movement could, and should, advocate the end of such a destructive system when capitalism is presumed robust, as much as when it is faltering.

The Worse, the Better

The idea that the worse things get, the better they will be for revolutionary prospects, dominates the other pole of catastrophist thinking on the left. Privations and hardship, the argument goes, push people to their breaking point—and from there, leftward. Economic crises and social immiseration commonly have been regarded as the necessary ingredients for social upheaval. State repression serves the same purpose—politicizing the apolitical and laying bare the system. The catastrophist take on both immiseration and state repression flattens into caricature what moves people to collective action, presupposing automatic reactions to dire straits. Both variants have led to wrongheaded politics—at times, tragically so. The notion of "the worse, the better" provides the starting point for several permutations that all flow from the voluntarist side of the catastrophist equation (although are sometimes imbued with a dose of determinism). As will be explored further on, it is the basis for the idea that if worsening conditions are more propitious for radical change, then radicals should do what they can to make things worse.

The axiom that worsening circumstances are favorable for radical change became a staple of various forms of socialist and anarchist revolutionary politics in the nineteenth century. The German-American evangelical communist Wilhelm Weitling believed that the more chaotic things became, the more auspicious they would be for revolution, and that chaos should be fomented by the most marginal members of society, such as robbers and bandits. An army of convicts would usher in revolution and institute a regimented barracks communism, or so he hoped.[35] Although Nikolai Chernyshevsky, the Russian populist writer, is reputed as first using the phrase "the worse, the better," its most notorious advocate was the Nihilist Sergei Nechaev—memorialized in Dostoevsky's novel *The Demons*, which depicted his instigation of a group killing of an ex-member of his secret organization. His *Catechism of a Revolutionist*—republished a hundred years later by the New Left to once again find an audience among contemporary insurrectionists—stated that his revolutionary organization "will employ its power and its resources in order to promote an intensification of those calamities and evils that must finally exhaust the patience of the people and drive it to a popular uprising."[36]

Marxism is frequently associated with the idea that worsening conditions—particularly destitution—engender revolution. Marx and Engels famously wrote in *The Communist Manifesto* that the immiseration of the working class under capitalism would intensify to the point of rebellion.[37] Yet in a work published the following year, they asserted something quite different: that revolt can emerge from rising expectations. "Although the enjoyments of the workers have risen," they contended in *Wage Labour and Capital*, "the social satisfaction that they give has fallen in comparison with the increased enjoyments of the capitalist, which are inaccessible to the worker, in comparison with the state of development of society in general. Our desires and pleasures spring from society; we measure them, therefore, by society and not by the objects which serve for their satisfaction. Because they are of a social nature, they are of a relative nature."[38] Of the two, the idea that the working class must be pushed to the edge of despair has often persisted within Marxism and the larger radical left.

A century ago, the anarcho-syndicalist Rudolf Rocker highlighted the problems with this type of thinking:

> The old slogan, "The worse the better" was based on an erroneous assumption. Like the other slogan, "All or nothing," which made many radicals oppose any improvement in the lot of workers, even when the workers demanded it, on the ground that it would distract the mind of the Proletariat, and turn it away from the road which leads to social emancipation. It is contrary to all the history and psychology; people who are not prepared to fight for the betterment of their living conditions are not likely to fight for social emancipation. Slogans of this kind are like a cancer in the revolutionary movement.[39]

Immiseration and eroding living standards do not automatically prompt workers to radical collective action. Workers find different ways to cope, some which would not win the approval of the left. Historically, workers have migrated in large numbers when threatened with impoverishment. They also often take actions, even collective ones, to shut other workers out of better jobs based on race, ethnicity, or gender—such as "hate strikes" by white workers against the hiring or promotion of workers of color.[40] Innumerable acts of solidarity and resistance, of course, mark the history of capitalism. But they are not the only recourse to which members of the working class resort in hard times.

Times of crisis can bring out the best in people, as a vast sociological literature on natural disasters has proved. Rebecca Solnit points out that while those in power summon up images of frenzied looters, pillaging and killing when calamity strikes—a fantasy that had deadly results during the aftermath of Hurricane Katrina in New Orleans, when poor black residents were demonized, abandoned, and murdered by the police and white vigilantes—most people pull together and engage in great acts of courage and mutual aid.[41] Yet as Solnit notes, "Leftists of a certain era liked to believe that the intensification of suffering produced revolution and was therefore to be desired or even encouraged; no such reliable formula ties social change to disaster or other suffering; calamities are at best openings through which a people may take power—or may lose the contest and be further subjugated."[42]

With the exception of the 1930s, periods of intense working class combativeness in the United States have tended to coincide with periods of economic expansion, not contraction and crisis. The two big strike waves of the early twentieth century, from 1898 to 1904 and 1916 to 1920, took place during years of growth. These were periods in which radical workers forced employers to raise wages—by 35 percent between 1890 and 1920—and, through struggle, successfully shortened the workweek by nine hours.[43] These strikes were fueled by relative prosperity, and industrial action fell off when the economy turned downward. "After 1910, each period of economic boom unleashed a rash of strikes by native craftsmen and immigrants alike," wrote labor historian David Montgomery. "Full employment, high turnover, unrest, and unions were blamed for an alleged decline in individual output of roughly one-fourth, which was especially evident among immigrant workers."[44] The big strike wave of 1916–1920 came to an end with the recession of 1920–1922, during which the strength of labor plummeted.[45] Even during the Great Depression, strikes, organized protest, and militancy did not significantly emerge until almost five years into the crisis, in 1933 and especially 1934; all were arguably fueled by a sense of hope that the world could be transformed, not one of absolute desperation.[46] Given the above, it should come as no surprise that the scapegoating of people of color—such as the demonization and deportation of immigrants—has historically increased during times of economic hardship. Lynchings of African Americans declined during the 1920s, but grew during the Great Depression (as did, it should be noted, opposition to them).[47]

The struggles and gains of the New Left—from the Civil Rights movement in the American South to the general strike in France in 1968—arose against the backdrop of rising expectations, driven by relatively affluent workers and students. One of the largest strike waves in the postwar United States erupted in the relative prosperity in the mid-1960s, "defensiveness giving way to aggressiveness and optimism," as social historian Cal Winslow describes it. A third of these strikes were wildcat—unsanctioned by the leadership of their unions. "The postwar years of near full employment and rising real wages had created confidence and combativeness among American workers. This confidence was fueled by the new politics of the 1960s."[48] The strike wave peaked in the years 1970 to 1974, along with U.S. wages, and both declined subsequently.

When capitalism experienced a major economic crisis in 1974, many radicals hailed it as finally furnishing the conditions for insurrection. The U.S. Maoist group Workers' Viewpoint concluded that more crises were on the way, crowing that the "1980s economic crisis will make the 1930s depression look like a picnic." Revolution, they reckoned, was imminent and preparations needed to be made forthwith for the dictatorship of the proletariat.[49] The strike wave of the 1970s declined, however, following the onset of the 1974 crisis. It came to an end in the recession of 1980, deliberately deepened by Federal Reserve chair Paul Volcker, along with Reagan's firing of striking air traffic controllers.[50] Rather than heralding a new era of radicalism, the crisis of the 1970s provided the opening for the right to amass power. Neoliberalism, not revolution, was the fruit born of capitalism's crisis and declining living standards for workers.[51]

It should be emphasized that I am not arguing that organized struggle only comes about during times of affluence—that is, that only improved circumstances lead to revolutionary change.[52] Rather, one cannot draw fixed and automatic conclusions from economic conditions. Social context, of course, shapes how people see their own situation and the forces at play around them, but there is no alignment of the stars that leads to collective, rather than atomized, resistance. Neither am I suggesting that radicals should fear crises, even if they embolden the right. Crises are an intrinsic part of capitalism—that much we can count on—and may provide openings for radical agitation. But they provide no automatic inducement for political consciousness and action.

* * *

On the radical left, the immiseration thesis has often been twinned with the argument that state repression creates the conditions in which revolution ripens and bursts forth. At times, leftists have hailed fascism and dictatorship as speeding the route to revolt. Others have welcomed increased repression by liberal democratic states as serving the same purpose. The most infamous example of this idea may be the Communist International's underestimation of the perils of fascism, overestimation of its "internal contradictions," and misreading of radical prospects following fascism. When Mussolini's fascists came to power in Italy in 1922, following a wave of strikes, factory occupations, and the establishment of workers' councils of the *bienno rosso*, some within the Italian Communist Party, or PCI, argued that fascism would be a passing phase to be replaced swiftly by socialist revolution, following from the stagist notion of the unfolding of history. As capitalism was expected to collapse under the weight of its contradictions, for some militants, so too would fascism fall apart.[53] When fascists gained momentum in Germany, the home of the largest Communist party outside of the Soviet Union, the response was similar. The party leadership unofficially adopted the slogan, "After Hitler, Our Turn." In 1928, Stalin via the Comintern had issued the assessment that after a period of revolutionary upheaval after World War I, and a period of serious setbacks and defeats in the 1920s, militants were entering a "Third Period" characterized by economic collapse and revolutionary militancy. In this new stage, revolution was a foregone conclusion and fascism a mere way station.

The German Communist Party (KPD), under the tutelage of Stalin, advanced the logic that the fascists would prepare the ground for revolution by making conditions worse. "The resistance of the masses to Fascism is bound to increase," wrote Heckert. "The establishment of an open Fascist dictatorship, by destroying all the democratic illusions among the masses and liberating them from the influence of Social Democracy, accelerates the rate of Germany's development toward proletarian revolution."[54] "By its adventuristic politics," resolved the Political Committee of the KPD, "fascism is pushing the internal contradictions . . . of German capitalism to the limit, and leading Germany to catastrophe. . . . So an immense revolutionary upsurge is beginning in Germany.[55]

The German Communist Party, blithely ignoring the danger of fascism, spurned the prospect of creating a united front with workers' parties against the Nazis. Given that capitalism was presumed terminally ill, fascism would not last and revolution was deemed imminent, no compromise was to be brooked with the Social Democrats. They were branded "social fascists" and were regarded on par with Hitler, if not worse. This led to the German Communist Party actually urging its members in the state of Prussia to vote for the Nazis in a referendum against the Social Democrats.[56] The ascent of the Nazis to power ought to have provided a mortal blow to the concept that catastrophic political and economic conditions inexorably lead down the road to radicalization and socialist revolution.[57] The Nazis promptly destroyed all worker and left-wing organizations and imprisoned their leaders—who provided the first inmates of the concentration camps—many of whom were executed.

* * *

While one might be dubious that state repression will induce revolution, an exception might be considered for situations of maximum violence: war. War has long been seen on the left as the midwife of revolt, often patterned in the trinity of crisis-war-revolution.[58] The connection between war and revolution is borne out by the two great waves of insurrection in the twentieth century, each following a world war, and the unraveling of the colonies following the latter conflagration. "The whole of Europe is filled with the spirit of revolution," opined British Prime Minister Lloyd George in 1919. "There is a deep sense not only of discontent, but of anger and revolt among the workmen against prewar conditions. The whole existing order in its political, social and economic aspects is questioned by the masses of the population from one end of Europe to the other."[59] Populations forced to go through the horror and atrocities of war tend not to be content with returning to the status quo ante. "Wars simultaneously push the lower classes to stand up and bow down," suggests Corey Robin, "giving elites more reasons to fear for their own power—and more instruments to advance it."[60]

As Beverly Silver has shown, in both the Global North and South, labor unrest—one measure of resistance and rebellion—hit historical highs immediately following the First and Second World Wars. Although labor militancy rose before both wars, it plummeted during most of the conflict. (Although radicals in World War I advocated militant resistance, the left as a whole fell into line with the nationalism promulgated by states.[61] People "pulled together" for the war effort, unified by nationalism not radicalism, whether because of patriotism or coercion. In many cases, workers received recognition of their unions or collective bargaining rights during the war in exchange for promises that they would not strike.[62] States also intensified their powers of compulsion.) In the final years, the horror of the war led to desertions and widespread anger. Returning soldiers had changed expectations about what they were owed and many were thoroughly disillusioned with the governments that had sent them to the charnel house. Veterans were adept with weaponry, while those who had shouldered the bulk of armaments manufacturing and production on the home front, including women, were not content to return to subordinate positions. (The omnipresence of weapons is a complicated affair. The German Revolution was initiated by sailors who mutinied when ordered into battle after the war was lost. Yet soldiers returning from the front were used to repress the revolutionaries and their workers' and soldiers' councils.)[63]

In the wake of World War I, revolutions shook the vanquished countries—Russia, Hungary, and Germany. But even the "victors" such as Italy were wracked with social unrest following the war. In the colonial world, revolutions and wars of emancipation have often followed wars involving the imperial nation—not simply because conditions worsened for the colonized, but because the colonizing power was weakened. India, which provided a million soldiers to the British during World War I, was on the verge of revolution at the war's end, although the Raj lasted two more decades thanks to the indecision of Gandhi and the Indian nationalists. World War II battered the large colonial empires—the French, British, and Dutch—while the Japanese empire was dismantled in defeat. "What fatally damaged the old colonialists was the proof that white men and their states could be defeated, shamefully and dishonorably," writes Eric Hobsbawm, "and the old colonial powers were patently too weak, even after a victorious war, to restore their old positions."[64]

If war has thus historically been linked to revolution, there is nothing inevitable about the relationship. War strengthens not only the left, but also the state—often in its most fearsome form. As always, the question is whether the forces of radicalism are strong enough and well organized enough, going into a moment of tumult, to prevail. The lessons of twentieth-century wars, decolonization, and imperial excursions have not been lost on those in power. As a consequence of war's destabilizing potential, they have significantly altered the ways they engage in combat. Since Vietnam, military personnel in the United States are no longer conscripted—a lesson learned from the wholesale refusal of ground troops to fight. Wars now depend more on high-tech weaponry, air wars, and drones (as in the NATO invasion of Afghanistan) than large numbers of boots on the ground.

Heightening the Contradictions

The idea that degraded conditions were more propitious for radicalism was widespread within the New Left and the many militant organizations that followed it in the 1970s. Markedly fewer activists actually followed this logic to its extreme: to make things more chaotic or bring down repression to hasten revolutionary change. A minority, however, believed that such contradictions could be intensified.[65] Having witnessed the radicalization of liberal activists after police beat them with truncheons, they jumped to the conclusion that the movement would grow if they could provoke police violence. At the 1968 Democratic National Convention in Chicago, Jerry Rubin unsuccessfully attempted to get fellow anarchist tricksters the Yippies to agree to "force a confrontation in which the establishment hits hard, thereby placing large numbers of people in a state of crisis and tension."[66]

Weathermen exemplified this tendency in the United States, emerging in 1969 when the New Left was caught in an impasse. Many were losing faith in the efficacy of peaceful mass demonstrations to stop the war in Vietnam. The Weathermen faction of Students for a Democratic Society took to the streets of Chicago to "bring the war home." While numbering far fewer activists than anticipated, they proceeded to battle the police and smash the windows of the city's wealthy district. Weatherwoman Susan Stern described their belief that by bringing repression down upon themselves, they would expose a fascistic system. "We weren't just a bunch of superviolent kids out to destroy Chicago because we enjoyed vandalism. . . . Mr. and Mrs. America would . . . see our bodies being blasted by shotguns, our terrified faces as we marched trembling but proud, to attack the armed might of the Nazi state of ours. Running blood, young, white human blood spilling and splattering all over the streets of Chicago for NBC and CBS to pick up in gory gory Technicolor."[67] Chicago Black Panther leader Fred Hampton disagreed with their strategy. He denounced them as "opportunistic, adventuristic, and Custeristic."[68]

In December of that year, the Weathermen met in Flint, Michigan for a "National War Council," at which they announced their intentions to go underground and to bring down the state through escalating chaos. Against this backdrop, U.S. empire was to be fought by Third World liberation movements externally and African Americans internally.[69] As member Cathy Wilkerson recalled, "The convention package given out to all who attended stated, 'Our strategy has to be geared toward forcing the disintegration of society, attacking at every level, from all directions and creating strategic "armed chaos" where there is now pig order.' I could see the logic behind the idea that our strongest weapon was our ability to create chaos. Chaos might weaken and eventually bring down the government from the inside, while third world countries extricated themselves from U.S. control. If we couldn't weaken the government, very likely we'd have a race war, sooner or later, I thought, so chaos now was better than a race-based chaos later on."[70]

In the aftermath of an accidental detonation of a bomb meant for military and civilian targets, the members of Weathermen rethought their direction and rejected the politics of "bringing on the bad."

The Weathermen were not alone in the 1960s and 1970s in believing that chaos would unmask the violence of the system and force open the eyes of the masses. In the affluent countries of the Global North, the term "fascism" was commonly used to describe liberal democratic regimes, with special resonance in postwar Germany, Italy, and Japan. Radicals believed that by making conditions worse, the true fascist nature of the state would be starkly revealed and the populations radicalized.[71] In Germany, where Nazism was only a generation removed, the Red Army Faction believed that they were facing a quasi-fascist regime hidden behind democratic window dressing—and that through targeted actions, radicals could force the state to drop its shroud.

"The enemy unmasks itself by its defensive maneuvers, by the system's reaction, by the counterrevolutionary escalation, by the transformation of the political state of emergency into a military state of emergency," wrote RAF leader Ulrike Meinhof in 1974. "This is how it shows its true face—and by its terrorism it provokes the masses to rise up against it, reinforcing the contradictions and making revolution inevitable."[72] She described how, two years prior, the actions of the RAF triggered a manhunt 150,000 police strong and led to the centralization of the police under the *Bundeskriminalamt*, the German equivalent of the FBI:

> This makes it clear that, already at that point, a numerically insignificant group of revolutionaries was all it took to set in motion all of the material and human resources of the state. It was already clear that the state's monopoly of violence had material limits, that their forces could be exhausted, that if, on the tactical level, imperialism is a beast that devours humans, on the strategic level it is a paper tiger.[73]

In more recent times, a politics has emerged that places contradiction-heightening at its center. Contemporary insurrectionism emerged out of the tumult of southern European radical politics of the 1960s and 1970s, although one could trace its lineage back much further. It forms two corresponding strands: anarchist and antistate communist. Anarchist insurrectionism originates from the work of Italian writer Alfredo Bonanno; the left communist variant has been shaped to varying degrees by the ultraleftism of Jean Barrot (*nom de plume* of Gilles Dauvé) and a farrago of Gilles Deleuze, Michel Foucault, Giorgio Agamben, and Alain Badiou. Anarchist insurrectionism is particularly influential in Italy, Greece, and Spain, while the antistate communist version has more currency in France and the United States. Over the past decade, insurrectionism has increased its appeal in North America, particularly following—and in reaction to—the disintegration of the antiglobalization movement. Its attraction in the United States was boosted partially by the left communist tract *The Coming Insurrection*, issued on the cusp of the new millennium by the French post-Situationist group the Invisible Committee. Repeated denunciations on Fox News by the right-wing commentator Glenn Beck made it an unlikely bestseller.

Insurrectionism in its anarchist form was shaped by the failure of large-scale radical social movements in Italy in the 1970s, and the advent of neoliberal restructuring of capitalism, with its just-in-time flexible production and transformation of the industrial working class, the classical radical agent of history. It eschewed the determinist notion that revolt depended on economic crises, and rejected the idea that one needed to wait until conditions—whether "objective" or "subjective"—were ripe for revolution.[74] "Why we are insurrectionalist anarchists," Bonanno explained, is "because rather than wait, we have decided to proceed to action, even if the time is not ripe. . . . Because we want to put an end to this state of affairs right away, rather than wait until conditions make its transformation possible."[75] (Echoing such sentiments, the Athens anarchist insurrectionist group Transgressio Legis approvingly cites Che Guevara's dictum that "It must always be assumed that all the appropriate conditions to launch the revolution are already in place. The outbreak of rebellion can make them appear.")[76] "It's useless to *wait*—for a breakthrough, for a revolution, the nuclear apocalypse or a social movement," declared the authors of *The Coming Insurrection*. "We are already situated *within* the collapse of a civilization. It is within this reality that we must choose sides. To no longer wait is, in one way or another, to enter into the logic of insurrection."[77]

Those insurrectionists who regard the "masses" as ultimately necessary for revolution—and those who think that a revolutionary rupture is in order—believe state repression will mobilize the unmobilized. In December 2008, the police murder of a fifteen-year-old boy sparked a three-week-long uprising in Greece. In the conflagration, insurrectionists saw the potential to enlarge the movement through state repression.[78] "When the rumors started to spread that the military would be called into the streets, we were wishing it would happen so that what we have been talking about for all these years, the civil war, the class war, would become a reality. . . . Our cause has always been to produce chaos," proclaimed Transgressio Legis.[79] "The State avoided a military-style operation, which would have catalyzed the arming of the rebellious crowd, and then we would all now be experiencing the beauty of the revolution."[80] To that end, they argue, small groups of people—organized informally without the presumed centralizing dangers of large-scale organizations—can force conflicts with the authorities that radicalize others through the tactic of attack, in the form of property destruction, battling the police, a bank bombing, and so on.[81]

There are many problems with radicals attempting to heighten the contradictions, not the least of which is bringing repression down on others for their own good. As a strategy, it lends itself to various forms of authoritarianism. It is a strategy, furthermore, that rarely goes according to plan. While increased repression may politicize some, others are likely to drop out of politics under such conditions, as was typical in 1970s when such strategies proliferated. Not a few New Leftists who fled from the cities to rural outposts at that time saw themselves as escaping the coming violence.[82] Moreover, it is a mistake to judge one's own strength on the degree to which repression is meted out. The state is often willing to use maximum force to repress even limited dissent. To base one's success simply off of police response is a sign of a movement's isolation. Swaggering expressions of hubris notwithstanding, insurrectionism builds not on the latent power of mass action, but rather on deeply held feelings of weakness.

Insurrectionism raises pertinent questions about how radicals should organize during times of ebbed mass struggle. But it answers them by making a virtue of temporary—even if protracted—periods. "One fact remains: we are isolated, which is not the same as saying we are few," an anonymous Italian insurrectionist anarchist tract declares.

> Not only does acting in small numbers not constitute a limit, it represents a totally different way of seeing social transformation. . . . To act when everyone advises waiting, when it is not possible to count on great followings, when you do not know beforehand whether you will get results or not, means one is already affirming what one is fighting for: a society *without measure*. This, then, is how action in small groups of people with affinity contains the most important of qualities—it is not mere tactical contrivance, but already contains the realization of one's goal.[83]

Or in Bonanno's words: "the revolutionary anarchist minority must be prepared for the historical task awaiting them."[84] While Bonanno and fellow insurrectionists criticized the vanguardism of professional revolutionaries like the Red Brigades, which they believed could be avoided by keeping their organizations small and informal, it is less clear that insurrectionism has escaped what it denounces. While Bonanno was hostile to workers' self-management in the 1970s, more recent iterations of insurrectionism have scathingly denounced as misguided "anarchist" consensus politics and other attempts at collective decision-making.

Undeniably, the state fears "chaos" or what it sees as social catastrophe—leveling impulses by the downtrodden against those at the top. Resistance and struggle can take many forms—the mob, the riot, the *jacquerie*, the uprising—and sometimes they can be chaotic and disorganized: Stonewall, Soweto, the Rodney King riots. "Decolonization, which sets out to change the order of the world," wrote Fanon in *The Wretched of the Earth*, "is clearly an agenda for total disorder."[85] "Order" as defined by capital and the state does not benefit the many. By definition, it is the status quo of accumulation, commodification, dispossession, racial and gender oppression, and empire. But the question remains: what kind of "disorder" and to what end? As other chapters in this volume illustrate, capital, the state, and the right (which is often but not entirely imbricated in capital and the state) also embrace catastrophe for their own purposes, and look to catastrophic redemption to wipe away the achievements of the subjugated. Out of the ashes . . . capitalism thrives.

The strategy of tension is a hallmark of the far right: sowing fear among the public to elicit a general clamor for law, order, security, and a strong authoritarian state to take matters in hand. From 1969 through 1980, Italian neofascists (with the complicity of parts of the military and intelligence services) bombed the country's trains, railway stations, demonstrations and public spaces, with the intent that the radical left would be blamed for the massacres and the public would demand an iron fist in response.[86]

Moments of crisis and strife are rife with dangers. The Mexican anarcho-syndicalist Ricardo Flores Magón presciently wrote in 1918:

> The insurrection of the peoples against existing conditions. . . . will be an overflow of indignation and bitterness, and it will produce chaos, chaos favorable to the prosperity of all of the fishermen in the river of revolt, chaos from which can surge new oppressions and new tyrannies, because in this chaos, regularly, the leader is the charlatan. Let us then, those who are conscious, prepare the popular mentality for when the moment arrives, because if we don't prepare for the insurrection it will give rise to tyranny.[87]

Social upheaval in the cause of revolution, as Flores Magón emphasized, was desirable. But social chaos would not necessarily occasion it.

Pining for the End Times

The outer edges of left catastrophism are inhabited by those who see the collapse of society—not simply capitalism, but civilization—as a route to a better world. One might say that the old slogan of a choice between "socialism or barbarism" has been upended, so that the latter is both inevitable and desirable. Both poles of catastrophism can be found here, at the edge of the precipice. For some, the collapse is preordained, the result of "peak oil," the scarcity of other natural resources, or the implosion of industrial society. For others, it needs hastening—even through nuclear war. Determinism and a determination to bring down the system through the actions of the few, as elsewhere, go hand-in-hand.

The most prominent proponents of this broad outlook label themselves "anticivilization" and often, but not exclusively, identify as anarchists. (Some embrace the name primitivist; others like bestselling author Derrick Jensen reject it.) While anticivilization ideas gained traction in the UK in the 1990s among animal rights activists and other environmentalists, spreading to the United States and Canada, it does not have much of a following outside of North America and Britain. Adversaries of civilization believe that the original catastrophe, begetting the catastrophe of the present, was the emergence of settled agriculture in the Neolithic period circa 8000 BC, where humans purportedly stopped living in harmony with the land and started organizing complex economies. A return to a hunter-gatherer society, they argue, is the only conceivable future that is sustainable. Such a transformation would necessarily require the dramatic reduction of the majority of the human population, since foraging would not provide a fraction of the food that the world's inhabitants need.[88] However—and auspiciously as they see it—the collapse may deliver such a reduction. As the now-defunct publication *Green Anarchy* puts it:

> Some time when you're on a busy street, in line at the post office, on the bus, look around. Get used to the idea that most of these people will not live a lot longer. Who among them would survive if the food stopped coming into the city for a

> month? . . . We will be throwing the stinking dead bodies of our families into pits and kneeling in garbage coughing up blood. But we may also get to break the pavement off the streets with sledge hammers and plant gardens. It's what's really going to happen: this civilization will fall, humans will survive, the Earth will survive, and we will have an opening to try something new. Within that range of imagined futures, even the bad extreme is not so bad, and at the good extreme we see the Earth quickly healing to its former fecundity, and people living peacefully with other life, and never sliding out of balance again.[89]

Opponents of civilization, if not yet apparent, tend to take a dim view of other humans. They eschew mass collective revolutionary action in favor of subversion by the chosen few who have little interest in whether others perish. "Revolution is not possible under the circumstances (nor desirable; insurrection on the other hand is)," writes Kevin Tucker, founder of the anarcho-primitivist journal *Species Traitor*, in celebration of chaos and crisis.

> We can no longer wait patiently for the "people" (whoever they are) to rise up against tyranny. Nor would I want six or however many billion humans coming together at once and destroying industrial civilization. . . .What can create crisis? What will impede or stop people from contributing to this death-machine? What will arrest progress (the evolution or trajectory of technoculture) dead in its tracks? Essentially, how can I bring down, cripple, or absolutely destroy my enemy? Not: How can I save the people? How can I intellectually force the dispossessed into accepting me as their vanguard? How can I be less alienating?[90]

Tucker sees revolutions as political, which is not a good thing, while the "primal war" that he advocates is about sabotage and returning to one's animal nature, or rewilding.[91] "I have no authoritarian vision (or desire for one) for ways of 'redistributing the wealth' or some other leftist pipe dream," he argues. "I see the fall of civilization to be inevitable, and thus, work to both brace for collapse and push for it, and for doing so I have no apologies."[92]

Derrick Jensen, Lierre Keith, and Aric McBay's tome *Deep Green Resistance* lays out a series of scenarios for taking down industrial civilization, including through the strategy of "decisive ecological warfare." (One critic described the book as an encounter between evangelical Protestant Noah Webster and the Red Army Faction.)[93] It predicts that the depletion of accessible oil reserves, or peak oil, will cause the global capitalist economy to fall apart around 2015. Large-scale manufacturing and agriculture will grind to a halt, governments will topple, authoritarian regimes will seize power and use mind control on their populations. Against this dystopian backdrop, radicals will have a choice either to sit back and watch, or to hasten the collapse, while organizing mutual aid through small autonomous communities. Decisive ecological warfare, the most militant of possible strategies, would aim to reduce fossil fuel consumption immediately by 90 percent through an escalating above and below ground strategy targeting industrial, and especially energy, infrastructure.[94]

> A drop in the human population is inevitable, and fewer people will die if collapse happens sooner. . . . Since unimpeded ecological collapse would kill off humans anyway, those species will ultimately have died for nothing, and the planet will take millions of years to recover. Therefore, those of us who care about the future of the planet have to dismantle the industrial energy infrastructure as rapidly as possible. We'll all have to deal with the social consequences as best we can. Besides, rapid collapse is ultimately good for humans—even if there is a partial die-off—because at least some people survive.[95]

Two generations earlier, the Argentine Trotskyist Juan Posadas of the modestly named breakaway Fourth International (Posadaist) took the idea of catastrophe ushering in radical change to its outer limits. He proclaimed that nuclear war was both inevitable and revolutionary, even though he acknowledged that it might wipe out half of the earth's population.[96] At the time of the Cuban Missile Crisis, Posadas encouraged the Soviet Union to bomb the United States, and later urged the Chinese government to do the same. "Capitalism hasn't ten years of life. If the workers' states launch support of the colonial revolution with all their forces, capitalism has not five years of life, and the atomic war will last a very short time," he declared.[97] "After destruction commences, the masses are going to emerge in all countries—in a short time, in a few hours. . . . [They] are going to come out, will have to come out, because it is the only way to survive, defeating the enemy."[98]

Posadas's ideas seem deranged. Indeed, six years later he argued that socialist aliens might liberate the earth from capitalism, a pinnacle in the politics of despair. (Ever forward-thinking, Posadas superseded the critique of socialism in one country, and opposed socialism on one planet.[99]) Yet he was not alone in his beliefs. Mao famously declared in Moscow in 1957 that nuclear war was not to be feared. "If the worse came to the worst and half of mankind died, the other half would remain, while imperialism would be razed to the ground, and the whole world would become socialist: in a number of years there would be 2.7 billion people again and definitely more."[100] Quite a few leftists in the 1960s and '70s shared the Great Helmsman's view that nuclear annihilation might wipe out the old order and allow a new radical society to be born.[101]

While the terrible prospect of nuclear war can never be ruled out, the chance of atomic annihilation clearing the way for socialism seems a bit farfetched. Likewise, while the threats of global warming, ocean acidification, and species extinctions are immense and horrific, it appears to be wishful thinking that "industrial society" will simply break down. Unfortunately for life on earth, there are enormous amounts of natural gas to be plundered from great swaths of the United States, filthy tar sands that could be processed, and profits to be reaped from renewable energies, once the price of petroleum gets high enough. Capitalism has shown itself destructively adept at vaulting obstacles to accumulation, and creating opportunities out of the misery of others.

* * *

The couplet of ideas that comprise left-wing catastrophism—that capitalism will collapse under its own weight or that worsening conditions automatically give rise to revolution—can be found across the left at various times and places. While one might conclude that Marxism tends toward the idea of predetermined collapse, and anarchism toward revolution regardless of conditions, this chapter suggests that such demarcations do not hold up. Radicals of all stripes—Maoists, left communists, populists, Stalinists, anarchists, radical environmentalists, Third Worldists, nihilists, and adherents of world-systems theory—have promulgated catastrophic politics, both fatalist and voluntarist. And anarchists, Marxists, syndicalists, and left communists have also opposed aspects of left-wing catastrophism over the last century and a half.

One might argue that what drives left catastrophism, on either side of the determinist-voluntarist dyad, is a certain maximalism—the uncompromising notion that revolution is always around the corner.[102] Depending on one's version of catastrophism, one either waits for the collapse to happen or summons revolution through the resolute action of determined cadre. In the latter version, any assertion that revolution might not be imminent, rather than being a realistic assessment of current conditions, indicates a lack of revolutionary will. (Radical lawyer Dan Siegel recounts this dynamic within a famed San Francisco Bay Area collective in the 1970s, partially comprised of *Ramparts* editor Robert Scheer and future California Senator Tom Hayden: "In the Red Family the debate was between those who thought the revolution would come in two years and those who said five. The two year group criticized the five year group as defeatists—my word—whose negative perspective justified a less than fully committed revolutionary life."[103]) Voluntarist catastrophism provides a shortcut for the urgent. One can sympathize with that haste, particularly in a time of colossal ecological crises. A certain degree of sloppiness, however, might also be detected. Why bother with understanding the conjuncture, engaging with the presumed deluded masses, when one can just take the purest, simplest, maximalist path and act? Though maximalism cuts across radical catastrophism, the latter can take less all-or-nothing forms as well. As we have noted elsewhere, versions of catastrophic ideas can be found among liberals. Hence, maximalism serves only to describe an aspect of left catastrophism, but is not a sufficient explanation for what engenders it and why it particularly proliferates in certain periods.

As Paul Mattick argued, Marxist crisis theory has ebbed and flowed along with the booms and busts of capitalism—interest peaking during periods of crisis, declining during periods of prosperity.[104] But left-wing catastrophism, in contrast, cannot be pegged simply to the fortunes of capital. While it seems to surge in times of economic crisis, it persists easily during times of expansion. As others have pointed out, the recurrence of the notion that capitalism will collapse has little to do with the theory's extremely poor predictive powers in the past.[105] Similarly, the idea of "the worse, the better" often flourishes during periods of relative prosperity, as in the United States and Europe in the 1960s when many leftists despaired of what they saw as an apathetic working class bought off by material abundance. While left-wing catastrophism is undoubtedly stoked in our time by justifiable fears of ecological collapse, economic mayhem, and the rhetoric of the state and right, the highs and lows of left catastrophism cannot simply be correlated to those fears.

Above all, what appears to account for the surges of left catastrophism, and propel its disparate forms, is a kind of despondent politics—what E.P. Thompson has called "the chiliasm of despair." While radical catastrophism can be found in times of upheaval and in times of moderate quiescence, it appears most pronounced during moments of relative impasse, when the forces of the left are divided or defeated. Thompson wrote in *The Making of the English Working Class* of the appeal of apocalyptic preachers, like Joanna Southcott, who amassed a large audience as prospects for revolution were extinguished in England, following the French Revolution and the onset of the Napoleonic Wars. Apocalypticism and a turn to Methodism were the "psychic consequences of the counterrevolution."[106] "He has lost all hope of paradise," Virginia Woolf wrote, "but he clings to the wider hope of eternal damnation."[107]

In retreat, emancipatory energies dissipate or become redirected. Having lost faith in themselves or others, those radicals remaining often hope that some force from outside might bring down a seemingly invincible system. It is this despairing element that allows left-wing catastrophism to take the often-contradictory forms that it does, alternating between mechanistic determinism and no-holds-barred voluntarism or adventurism—often within the same individual or movement. Voluntarism inverts despair, with an isolated and small number of committed activists acting on the moral imperative to create change regardless of the limits of the possible: vanguardists proclaiming the one true path, righteous activists deciding that they are the "exceptional white people" that understand empire and racism, insurrectionists hoping the police will crack down harder so that they can proclaim a "civil war." Both faces of catastrophism lack faith in the power of mass action and require either external forces or the leadership of the select few.[108]

It ought to be noted that some proponents of catastrophic politics on the left have been driven less by despair—whatever their lack of faith in mass action and collective democracy—and more by baser motives. During the height of the Great Proletarian Cultural Revolution, when the state displaced large numbers of urban dwellers to the countryside, persecuting millions more, the Chinese *People's Daily* emblazoned Mao's words on its front page: "Great Chaos under Heaven—the Situation is Excellent." One imagines that foremost in Mao's mind was a desire to reassert power, not paroxysms of despondency. Likewise, undoubtedly some on the voluntarist edge of the catastrophist spectrum arrived at their politics out of hubris and an elitism of sorts. The allure of being one of the chosen, seeing things as they really are, has persisted through generations of movements.

After almost four decades of radical retreat, it is no surprise that left-wing catastrophism is so pervasive in our era. The liberatory hopes of the past, and the confidence in the collective power of others has given way to the uncertain hope and fear of collapse, befitting our anti-utopian and crisis-fraught times. Even in the darkest of hours, however, it behooves radicals to construct a politics that categorically rejects catastrophism. Theodor Adorno, no stranger to pessimism, warned of the undialectical nature of seeing the world in such grim terms that only an exterior force could change it.[109] That is a sentiment we should heed.

While the underlying motive of catastrophism—and hopes for sweeping change embodied within it—is entirely understandable, neither fatalism nor voluntarism usefully serve a radical emancipatory project. On the contrary, they often do great harm to its prospects. No amount of fire and brimstone can substitute for the often-protracted, difficult, and frequently unrewarding work of building radical mass movements, even under situations of the utmost urgency. When they deploy catastrophic rhetoric, radicals overlook the diminishing returns and distorting effects it has on the forms of organizing that it does manage to inspire. Fear is corrosive. It is especially corrosive within the left. The right thrives on fear, while on the left it mobilizes only a core few. One need only think about the Revolutionary Communist Party slogan of the mid-1980s: "A horrible end, or an end to the horror?" Why bother joining a movement if the end is nigh? Radical mass movements typically grow because they offer hope for positive change, while fear demobilizes.

Navigating away from the stormy shoals of catastrophism, between the Scylla of voluntarism and the Charybdis of fatalism, requires a commitment to mass radical collective politics, in inauspicious times as well as auspicious ones. Tectonic upheavals can burst out of such endeavors, often quite unpredictably. It necessitates rejecting the heavy-handed politics of the chosen, and the quietist politics of the armchair determinist, waiting for iron laws of history to bring an end to the capitalist order. If we are committed to the demise of capitalism, we should steer well clear of catastrophism.

CHAPTER THREE

At War with the Future: Catastrophism and the Right

James Davis

O strange men!
That can such sweet use make of what they hate.
—Shakespeare

THE AMERICAN BROADCASTER LOU DOBBS CREATED A FIRESTORM OF controversy when he reported that illegal immigrants crossing the U.S.-Mexico border were introducing leprosy into the country and "threatening the health of many Americans." His reporter claimed that seven thousand cases of leprosy had been recorded in the previous three years, a number that was quickly established by critics to be wildly inflated. Dobbs stood by the report, though, and defended it on many subsequent occasions. "The fact that we're somehow helpless to defend this country, to secure our borders" was the point being made in the segment and illegal lepers sneaking through the borderlands was mere illustration.[1] The leprosy claim is one among many negative stereotypes and slurs that the right associates with the undocumented in the United States and uses to feed broader fears about White Judeo-Christian America being under siege. Over the past two decades, organizations like the California Coalition for Immigration Reform and later the Minutemen Project have succeeded in generating hysteria around illegal immigration, helping justify the creation of the U.S. Immigration and Customs Enforcement agency (ICE).[2]

The border has become its own political crisis, with the deaths of thousands of migrants pushed to cross more treacherous desert areas due to a border fence, or at the hands of vigilantes, or Border Patrol agents.[3] Dobbs's hyperbole was a footnote in this larger context of border policy in recent decades but catastrophist panic is part of the politics that have helped formulate the policy, even if mainstream critics of Dobbs failed to acknowledge the connection.[4]

Catastrophism—defined as a political orientation premised on the assumption that society is on course for an economic, environmental, social, or spiritual collapse due to forces internal or external to us, out of which a new society may emerge—is central to the propaganda and ideology of the modern right.[5] Catastrophe itself takes on two forms in right-wing imagination. The first is that catastrophe is the inevitable ongoing result of any gains by the left. This universal conviction emerged as part of right-wing reaction in the decades following the French Revolution. For the right, the political advancement of previously subjugated groups and classes of people represents an immediate threat to privilege and status, but taken to a revolutionary extreme it also represents an absolute threat to the existence of the ruling class. Such advances are deemed catastrophic for that existing order. As a consequence, Spengler's "twilight" of the West haunts the political right.

The second version of this ideology expounds a catastrophist antidote whereby enemies are confronted and vanquished in a final apocalyptic conflagration through race war, insurrection, Armageddon, civil war, or in its purest form, biblical apocalypse, and the rapture. This belief is often associated with the idea that apocalypse can be overcome and survived and that a purer form of existence and society is possible. This notion of rebirth contrasts with the prevailing view of catastrophe as a disease of social movements and creaking hierarchy. Here, catastrophe is the cure for a decadent social, moral, and political order.

It is impossible to understand the catastrophist extraparliamentary right without analyzing its relationship with the capitalist state. While the right often views the state as the mechanism through which the left accomplishes its radical project to dismantle the "traditional" order, the right also exercises considerable influence over the state and state policy. In turn, the state appreciates and exploits the organized right as defender of the status quo, and gains political space from its success.[6] By intensifying paranoia and division about immigrants, welfare, external and internal security threats, fiscal crises, morality, and minorities, the organized right works to generate a climate in which the state can "react" to various supposed crises. Indeed, the right often succeeds in generating genuine political crises through agitation and propaganda alone. Border security is just one example where agitation from the right has contributed to the state's militarization of the U.S.-Mexico border, and the growth of a massive internal security apparatus.[7]

The state, too, has its own forms of catastrophism that serve to provide an outlet for the emergency operation of the modern capitalist security regime. In order to expand its own power internally and externally, the state uses the prospect of impending catastrophe to dismantle popular opposition to war and to justify the expansion of state power over its own people. The march to war has been accompanied by the manipulation of fear and the ominous spectre of existential threat. In practice these threats are exaggerated to make war look like a reasonable form of defense, or the only means of survival. "We don't want the smoking gun to be a mushroom cloud," as then-U.S. National Security Advisor Condoleezza Rice put it.[8] While it is extremely rare for an advanced society to disappear suddenly by virtue of some extreme calamity, it is a strikingly common form of propaganda deployed by those in a position to control events.

This chapter takes a brief survey of right-wing catastrophe from the prebiblical roots of the apocalyptic idea through its subsequent adoption by right-wing ideology and to the catastrophist manipulation of real and conjured fears by the modern nation state. General anxieties about death, calamity, and vulnerability are as old as literature and probably as old as storytelling. Catastrophism enters human history inside this narrative form, and has been carried into the modern world primarily through religious belief and tradition. The right has adapted the narrative model of religious apocalypse for its own propaganda, and it has extrapolated a secular form of catastrophic prediction and analysis with it and applied it here on earth. Catastrophism is a successful propaganda weapon and a recurring component of right-wing organizing logic.[9]

This binary of catastrophe as the disease and/or the cure is useful to model how the right employs the rhetoric of catastrophe and it structures this discussion. Disease catastrophe, or that which sees the advances made by social movements of the left as catastrophic, is universal in right-wing ideology. By definition, the right is at least partly a reaction to democratic expansion. Cure catastrophists additionally believe that a catastrophic remedy to the disease of catastrophe is warranted and that such a development should be welcomed and even fomented. This essay first will examine right-wing catastrophe as the cure and then look at the broader right-wing politics of catastrophe as the disease. Finally, it will examine how the state uses right-wing catastrophism to buttress its own interests.

Catastrophe as the Cure

The Christian idea of the apocalypse generally refers to the second coming of Christ and the final judgment of everyone by God. It is understood as both the revelation of truth and as the end of history. It is also the source of the secular idea of catastrophe as collapse and rebirth. Evangelical Christians whose worldview is definitively shaped by religious apocalypse believe that the end-times catastrophe will overwhelm evil and restore only the virtuous to God and heaven. Here, we will trace cure catastrophism from its roots in the religious apocalypse through its secular turn during the Cold War and up to its modern secular expression by adherents like the Norwegian far-right mass murderer Anders Breivik.

In the early nineteenth century, John Nelson Darby developed a system of belief known as "dispensational premillennialism." In it, he elaborately articulated a stages theory of history in man's relationship with the Almighty and divided scripture into epochs. Ultimately, the various destinies of Jews and Gentiles would unfold in the last days before rapture, when "believers" are transported to heaven before the apocalypse. Darby was a strict Calvinist and preached the absolute sovereignty of God with little opportunity for human agency. "The consequence of sin is not ceasing to exist—it is death, and after that, judgment."[10] Dispensationalism precisely accounts for the temporal unfolding of the end of history: the Jewish return to Israel; the emergence of the "beast" and the great tribulation; and the return of Christ with an army of saints to vanquish the combined might of the neo-Roman empire, the false prophet and the beast at a site in Armageddon, located at Megiddo in northern Israel. Darby's apocalypticism is a foundational model of cure catastrophism, establishing a purifying conflagration that assures utopia for the few and the destruction of those who ignore their prophecy.

In the latter part of the nineteenth century, a network of evangelical bible schools and conferences trained thousands of ministers and lay people to carry forth this fundamentalist apocalyptic vision and prophecy throughout the United States. Cyrus R. Scofield was one of those it reached. A Confederate veteran of the Civil War who found salvation languishing in a St. Louis jail on forgery charges, Scofield came under the mentorship of James Brooks, an early Darby enthusiast who introduced him to Darby's interpretations. Scofield was an energetic and able convert who rose to prominence within the movement as a speaker and author.[11] In 1909, Oxford University Press published his *Scofield Reference Bible*, closely annotated with dispensationalist interpretations of the text. It sold more than ten million copies and has become one of the most popular sources for millions of evangelicals and fundamentalists over the last hundred years. Now in the public domain, it remains a bestseller and describes many of the convictions associated with rapture and the end of days.[12]

Premillennialists believe that, following the second coming, Christ will reign for a thousand years. Those influenced by Darby's dispensational index asserted that earthly progress was illusion, that the world was becoming inexorably more wicked and that mankind's efforts at reversing it were in vain. From the latter part of the century until the World War I, even some explicitly premillennial organizations displayed a firm commitment to charity and even to social reform.[13] Social programs and charity were widely supported, but disagreements arose over whether they took private or public form. Nevertheless, most nineteenth-century evangelicals regarded the state as a legitimate means of restraining evil. But with the profound changes that this period wrought on American society (city slums, industrialization, non-Protestant immigration, political liberalism, and ultimately World War I), these tensions became increasingly difficult to manage. Within Protestant denominations, believers in the literal truth of the Bible battled with believers in the "Social Gospel," and a status quo in favor of tolerance for theological diversity was established by the end of the first decade of the twentieth century. But by 1920, social action and political reformism had all but disappeared from the evangelical scene. In many respects, modernity resolved the internal conflicts of the evangelical Protestant Gospel by presenting various cultural threats to traditional Christian hegemony.

By the 1920s, evangelicals were also confronted with a liberalizing cultural landscape loosely encapsulated by the Jazz Age. It strengthened a pessimistic cultural view among fundamentalists who saw evidence of debauched and hopeless trends in the secularizing world. The 1920 census recorded an urban majority for the first time. By the 1930s, a broad political culture began to take shape within fundamentalism. Anticommunism was already prevalent among religious people and the Russian Revolution was firmly associated with atheism and deicide. The New Deal convinced many of them that Roosevelt was "painting America red." Apocalyptic and prophetic literature is rich in sensational accounts of Satan's plots against the church and the faithful, and the 1930s was a high point of "beast watching." Fundamentalist writers and preachers saw the Antichrist in the rise of fascism and communism, but particularly in Mussolini's links with Rome and the Pope.[14]

While the Antichrist was traditionally believed to be the Pope, and the Catholic Church had long been regarded as "a monstrous beast," changing social and political conditions regularly threw up new candidates.[15] During the War of Independence, the British Empire featured in sermons as the likely Angel of Death devoted to undermining America's liberties and Christian faith.

Subsequently, the anticlerical Bavarian Order of the Illuminati was identified with the French Revolution and with plots to overthrow European governments. Timothy Dwight, the president of Yale, introduced the conspiracy to the United States and for him the Antichrist was unmistakably associated with the French Revolution. "Shall our sons become the disciples of Voltaire, and the dragoons of Marat; or our daughters the concubines of the Illuminati?[16] Both the anti-Christ and the Illuminati offer a blank canvas for conspiratorial paranoia about internal and external enemies that resist rational interrogation. Currently, the New World Order, or a plot by global economic and political elites to impose a single world dictatorship, provides a screen onto which multiple conspiracies can be projected. Each can be invested with the most extreme malice, yet the outlines and effects of these conspiracies remain just over the horizon, properly understood only by an informed elite.

A fascination with the illuminati remains a popular motif of American catastrophic conspiracism. In the twentieth century, the dispensationalist writer Arno C. Gaebelian found a wide-ranging set of conspiracies within and emanating from Soviet communism and linked it back to Dwight's obsession, the Illuminati. Gaebelian characterized Russian Communism as an important part of Satan's project to destroy America and led an evangelical assault on "the so-called New Deal, with its colossal failures, [that] has in it a good deal of the socialistic-Communistic theories . . . traceable to certain 'brain-trusters' who were Moscow inspired and Moscow directed."[17] Conspiracies involve secret cabals of enemies who follow a script that ends with the loss of civilization or liberty and that can only be resisted by the heroic and lonely efforts of an informed elite. The Illuminati conspiracy remains a potent mania on the right and is repeatedly implicated in the New World Order plot, much as it was in the French Revolution.

Godless Communism

Shortly after the Russian Revolution, an apocalyptic doomsday politics that combined biblical prophecy with secular Cold War catastrophism developed on the radical right, who thus viewed

Soviet communism as a mortal threat. The establishment of the state of Israel in 1948, realizing the prophesied Jewish return, enhanced a mood that the end times were at hand. The essential outline of the Cold War itself folded anti-communism and nuclear Armageddon into a satanic miasma of apocalyptic promises: "The . . . anti-Christ stalks our world. Organized communism seeks even to dethrone God from his central place in the Universe."[18] Democratic Senator and presidential candidate Adlai Stevenson, expanded on the National Security Council's warning that "the Soviet Union . . . is animated by a new fanatic faith . . . and seeks to impose its absolute authority over the rest of the world."[19] Public utterances of faith are stock in trade for all types of politicians and U.S. tropes of freedom and liberty are closely associated with morality and religion, as the popular view of history is colored by scripture and even secular histories emphasize exceptionalism and manifest destiny rather than any continuity with the past. Moreover, social and economic norms render experiences of mobility and materialism that challenge identities established around tradition, community, or class. The scriptural idea of the millennial kingdom to come is projected onto the place and many people understand divine providence as part of the American story. As such, the idea of divine superintendence over America is definitively tied to biblical apocalypse, with America itself as the postapocalyptic nirvana.

The Cold War created a social and political laboratory in which the secular American right and the evangelical movement could find a common set of obsessions. The coincidence of interests between press barons Henry Luce, owner of *Time* magazine, and William Randolph Hearst on the one hand, and a preacher like Billy Graham on the other, was an early union of right-wing politics and apocalyptic believing that would grow into a rich collaboration decades later. "I believe today that the battle is between communism and Christianity . . . We need a revival!"[20] Graham exploded into national consciousness through his revival meetings in Los Angeles in the late 1940s. Heavily promoted by Hearst and Luce, Graham consolidated his preeminence among American Protestants. He became one of the most admired men in America and attracted enormous crowds even in Europe. Graham expertly wound scriptural apocalypse with general anxieties about the decline of America's Christian identity, while appealing to both Hearst's and Luce's concern over Soviet communism and social democracy.[21] More than two million people saw him in London during three months in 1954. "I found that scripture texts became like a rapier in my hand."[22] As the 1950s unfolded Graham molded for a refreshed white middle class an identity that fused patriotic and apocalyptic religious fever with consumerism and leisure pursuits.

But internally, threats to the established order were also gaining a momentum that would challenge the racial and gender status quo and fuel a catastrophist backlash from the right. For most of the twentieth century, the United States was still an unarguably Protestant country. For example, opposition to Catholic Al Smith's presidential candidacy in 1928 ran across liberal and conservative Protestant lines, as did support for prohibition. At the time, there was little consensus for any Protestant politics beyond sectarian prejudice directed at Catholics and sometimes Jews. Southern fundamentalism had traditionally focused on piety and the private nature of the struggle for grace, eschewing direct political activity. Typically white evangelicals had enjoyed cultural dominance throughout the South and while that lasted there was little need for church directed political organizing as they were rarely in conflict with local political orthodoxies. But millions of southerners living in the North and West of the country found themselves involved in increasingly contentious disputes. The 1963 Supreme Court decision that ended school prayer and the introduction of sex education in schools later that decade mobilized Christian fundamentalists and accelerated their political engagement. Sex education in particular was seen as a direct assault on the family. The Civil Rights movement by the 1950s began to pressure the racial and social order and threatened white evangelical supremacy in the South. Combined with an increasingly permissive regime in areas of gender and sexuality, and crescendoing with the legalization of abortion in California shortly after Reagan's election to the governorship in 1967, evangelicals were drawn further into a fundamentalist backlash. Mass TV communicators like Jerry Falwell and Pat Robertson, who owned the first private communications satellite, contributed to the growing coherence of the movement and to increasing alarm over gathering threats. The emergence of social tolerance and the partial breakdown of race and sex hierarchies channeled apocalyptic religious divination into the social and political sphere and helped codify a political catastrophism that could hide some of the seeming contradictions between religious apocalypticism and the propagandistic catastrophism of the cultural right.

The New Old World

The right sees the democratic expansion achieved by Civil Rights and feminism as the result of a Marxist conspiratorial force rather than of social movements that are historical actors in themselves. Similarly, liberal and left support for immigration and multiculturalism is regarded by cure catastrophists as a conscious attack on traditional demographic and political identity. For the mainstream, this is often a purely electoral calculation that both distracts from the pro-business hegemony they promote and destabilizes the left opposition. But on the extreme right, many of these same ideas justify a homicidal hatred of the left that was brutally exemplified by Anders Breivik when he murdered seventy-eight people at a social democratic youth camp and in downtown Oslo in 2010.

There is an odd fashion among some young Nordic men to own and drive 1950s era American cars, Chryslers, Buicks, Oldsmobiles, and the like. It's possible that this scene appealed to Breivik and that he admired the muscular aesthetic of the cars and their lacquered drivers as they cruised downtown on a Saturday night. He quotes (without attribution) William Lind from the American Free Congress Foundation, "Most Europeans look back on the 1950s as a good time. Our homes were safe . . . children grew up in two-parent households, and the mother was there to meet the child when he came home from school." Early on in his "compendium," he credits a world of orthodox securities to this Rockwellian idyll that probably corresponds more closely with the lives of some of his Muslim neighbors than those of his post-Christian compatriots he seeks to influence. In the intervening period, however, the family's foundational role in Western Europe has been diminished, along with general attitudes toward morality, authority, gender discipline, and Western Christian values. Following Lind's lead, Breivik traces these changes back to the Frankfurt School and he argues that Georg Lukács's formulation of "the archaic nature of . . . bourgeois family codes, the out-datedness of monogamy, and the irrelevance of religion," launched "cultural Marxism" or the translation of Marxism from economics to culture.[23]

The Frankfurt School conspiracy theory promotes the idea that a small group of European Jewish intellectuals who emigrated to the United States in the 1930s unleashed a program intended to undermine Western Christian values and deliver the American people into the hands of the far left. According to Breivik, Lukács, Wilhelm Reich, Theodor Adorno, and their followers regarded the Western individual as a product of "Christianity, capitalism, authority, the family, patriarchy, hierarchy, morality, tradition, sexual restraint, loyalty, patriotism, nationalism, heredity, ethnocentrism, convention and conservatism."[24] If Lukács opened a front against the family, Adorno's *The Authoritarian Personality* attacked the individual and is the *über* text from which springs "political correctness," "a radical inversion of the traditional culture in order to create social revolution" and a totalitarian assault on authority and on traditional values. Herbert Marcuse then bridged the generation gap between the Frankfurt School and the 1960s counterculture, where Western youth "burned the flag and took over college administration buildings."

The Frankfurt School conspiracy theory embraced by Breivik, along with Pat Buchanan, David Horowitz, William Lind, David Duke, Paul Weyrich, and many others on the U.S. right, was first advanced by Michael Minnicino in *Fidelio* magazine in 1992.[25] According to Minnicino, "a new Dark Age is exactly what we are in. In such situations, the record of history is unequivocal: either we create a Renaissance—a rebirth of the fundamental principles upon which civilization originated—or, our civilization dies."[26] *Fidelio* was a publication of the Lyndon LaRouche network/cult and Minnicino's theory was picked up by Paul Weyrich and the Free Congress Foundation, ultimately resulting in a documentary film outlining the conspiracy called *Political Correctness: The Frankfurt School*.[27] The conspiracy sets out an elaborate scheme to destroy Western Christian civilization initiated by Lukács during his time serving in the short-lived Hungarian Soviet Republic government. The film features Lázló Pásztor who had been a youth leader of the Hungarian Arrow Cross, and served in the Nazi supported Hungarian government during World War II. Pásztor went on to serve as the first Chairman of the Republican Heritage Groups Council and was close associate of Paul Weyrich, one of the founders of the Moral Majority and a hugely influential figure on the American right who died in 2008.[28]

The Frankfurt School's real genesis was an effort to come to terms with the rise of fascism inside the most advanced economy in Europe during the 1930s. Lukács, Sigmund Freud, Ernst Bloch, and countless others lived through and concerned themselves with fascism. Germany had the largest Communist Party, the most successful Social Democratic Party, and the most powerful labor movement in Europe yet all collapsed in face of Hitler and the Brownshirts. Max Horkheimer pondered "why humanity, instead of entering a truly human state, is sinking into a new kind of barbarism" and his inquiries gave birth to Critical Theory. One reason that the Frankfurt School is of interest to the far right is that its analysis closes off the idea that fascism is some sort of aberration in capitalism; instead, it is part of their contribution to argue that fascism is a logical extension of capitalism *in extremis*. Most neo-fascists are squeamish about their own political lineage; indeed contemporary figures like Glenn Beck often use "fascism" as a swear word. But the Frankfurt School described the association of capitalism with fascism and suggested that the far right is an authoritarian instrument for defending capitalism under duress. Another reason the contemporary right still battles with the legacy of the Frankfurt School is that its ideas also appeal to an audience beyond that available to those pushing narrow class politics. "Cultural Marxism" is hated by the right because it influences middle class students who go on to positions of influence. For the right, pro-business arguments do not work as effectively against a critique of alienation and mass culture as charges of Soviet tyranny might have done in disputes with orthodox Marxism.

As far as Breivik was concerned, the young social democrats and socialists he murdered were in fact child soldiers in the Frankfurt School's revolutionary army fighting to destroy "the West." For him, political correctness now dominates Western society to such a degree that it undermines freedom of speech and academic freedom. Smug feelings have replaced critical thinking and political indoctrination is what universities do. He echoes Richard Nixon's 1968 warning that the Columbia University strike signaled the emergence of a "revolutionary struggle to seize the universities of this country and transform them into sanctuaries for radicals." Meanwhile, the European male has been emasculated by feminism and political correctness and rendered unable to defend traditional beliefs. Multiculturalism, diversity, and affirmative action are just a few of the tactics used by cultural Marxism to prosecute the destruction of European Christian civilization. "Leftists who dislike Western civilization use multiculturalism to undermine it, a hate ideology disguised as tolerance," writes Breivik. "Multiculturalism equals the unilateral destruction of Western culture, the only unilateral action the West is allowed to take."[29]

Clearly, there are many on the right who would not go so far as Breivik to try to instigate a violent transformational catastrophe. Nevertheless, they share his conviction that a putative contemporary status quo of feminism, multiculturalism, and the enduring vestiges of social democracy are in themselves catastrophic. The next section explores this broader impulse, which is near universal on the right, that the victories of the left are catastrophic.

Catastrophe as the Disease

Across the American and European right, there is a persistent theme that Western civilization is afflicted and decaying. At the ideological root of this preoccupation is recognition that the twentieth century has been defined by the gains of the left. Feminist struggles for gender equality have perturbed and provoked the right for more than a century. Suffrage, labor market participation, reproductive freedom, subsidized child care, sex education, and gender-related social movements are key questions around which the right has organized to defend patriarchy. The formal decline of white supremacy and the partial disappearance of legal racial apartheid is another source of grievance around which right-wing politics orbits. Alongside these, social democracy and trade unionism are just a few of the historical traditions that have profoundly altered the distribution of power within American and European societies. Disease catastrophe, which is a much larger category than cure catastrophe, knows that these victories cannot be easily undone and that their achievement has been to permanently alter how privilege and authority is distributed throughout society. Continuing right-wing attacks on reproductive rights, multiculturalism, and the ability of workers to organize and bargain collectively are efforts to reverse these advances, or at least check their momentum. Framing them as grave threats through fear and scapegoating invites extreme remedies from the state and encourages mainstream politics to view and discuss these issues in terms of social breakdown and calamity. This is one way in which right-wing catastrophism impacts the policies of more moderate conservatives.

Related to this antagonism toward the political achievements of social movements is a highly unrealistic view of what the right has actually accomplished in the last thirty years. Despite the ascendancy of neoliberal capitalism and the accompanying decline in the economic security of workers, the gutting of the welfare state, and the incremental fall in taxes owed by the rich and by corporations, the free market right nevertheless feels it has lost. This pessimistic view is useful inasmuch as it sustains a siege mentality on the right and reinforces the idea that an alien social transformation is imminent. It is difficult to foresee a total reversal of the advancement of the position of women before the law and in the economy, for example, but it is possible to imagine that their position might be further enhanced. If the object of democratic social movements is the establishment of equality between all people then, as far as the right is concerned, the war will continue.

From the point of view of the Austrian School of economics and the capitalist right, catastrophe is literally social democracy, recognized as a mortal threat to free market hegemony. Friedrich Hayek conjures its opposite: "If we ask what men most owe to the moral practices of those who are called capitalists the answer is: their very lives." We are to believe that humanity is the historical invention of capitalism *ex nihilo*, that the proletariat has been produced whole by the "moral practices" of capitalists alone, and that markets deserve a noble sort of sovereignty.[30] The capitalist right remembers the gilded age of nineteenth-century *laissez-faire* capitalism as both the golden years and the Promised Land. Ayn Rand has it as a lost utopia in her totalitarian purgatory where superman decries and confounds the central planners.[31]

For disease catastrophist Leo Strauss, the twentieth century was an era of "permissive egalitarianism" in which virtue, understood in the way that was valued by classical Greeks, is lost in the headlong rush for a relativistic modernity.[32] Permissive egalitarianism is Enlightenment-inspired democratic expansion and "the foremost duty of political philosophy to-day seems to be to counteract this modern utopianism" which is "bound to lead to disaster."[33] Strauss focused his philosophical work on the Classics but his followers divined from it key insights into the modern world and he remains an important inspiration on the American right. Reagan's formulation of the Soviet Union as the "evil empire" owes something to it, as does the duality established in the war on terror between good and evildoers. Strauss had an important early relationship with Nazi jurist Carl Schmitt and was drawn to Schmitt's Hobbesian pessimism. Strauss observes that for Schmitt "the life of man gets its specific political tension from the possibility of war, from the dire emergency, from the extreme possibility."[34]

The idea of egalitarian fallacy is the thread that unifies the right before and after Strauss. Deeply at odds with the democratic shibboleths of modernity and with the religious notion of equality before an omnipotent deity, thinkers from Hobbes and Burke to Von Mises and Rand have railed against democracy and celebrated the outstanding individual swimming against the mediocre tide. Von Mises elaborates a secret history of humanity whereby all advance is the result of the contribution of a small few exceptional characters who stand above and apart, and without whom humanity would exist in a cannibalistic twilight where hell is other people. Writing about Nietzsche for an American audience, H.L. Mencken declared that "only the underdog believes in equality. . . . It is only the groveling and inefficient mob that seeks to reduce all humanity to one dead level, for it is only the mob that would gain by such leveling."[35] Ayn Rand created a penny literary cult from all this that appeals to broad sections of bourgeois society in which catastrophe is played by the masses arriving to claim their place in politics and history.

Strauss's influence on the neoconservatives is evident in their ambivalence to a purely economistic evaluation of the world and their antipathy to an obsession with free markets as an endgame of suitable epicness. The idea that the invasion of Iraq was conceived by neocons exclusively to allow U.S. corporations to prosper fails to appreciate that their motivation is more about hegemony than about markets and capitalism. Their enemy is inertia as much as it is their foes. Reagan's crusading war with the evil empire was more like it, as was the openly imperial role that the U.S. seemed destined to play after 9/11. Reagan's triumph in 1980 was the result of tectonic shifts in American politics and demographics that have their roots in the 1960s. By 1970, more than seven million transplanted white southerners lived outside the South and in a process that stretched back to the 1930s they could be found in fundamentalist churches throughout the North and particularly in the Midwest and Southern California. Billy Graham came to prominence in LA out of this milieu. The transformation of American politics that occurred following Lyndon Johnson's "loss of the South" to the Republicans following the Civil Rights Act reiterated race as the defining catastrophic domestic political question for the country.

As with culture fixated cure catastrophists like Breivik, for many right-wing disease catastrophists, the roots of decay lie in multiculturalism and cultural relativism, or the idea that significant differences between groups are cultural rather than biological. The threat posed by multiculturalism is seen as both the responsibility and result of liberalism and social democracy. Right-wing nationalists in the United States identify a race catastrophe resulting from the browning of America through integration and immigration from the South. *Mexifornia* author Victor Hansen, for example, draws parallels between U.S. immigration and the immigration of Muslims in Europe. "The riots in France, the support for jihadism among Pakistanis in London, and the demands of Islamists in Scandinavia, Germany, and the Netherlands do not encourage Americans to let in more poor Mexican illegal immigrants with loud agendas, or to embrace the multicultural salad bowl over their own distinctive melting pot."[36] Hansen locates the engine of multicultural promotion in the work of academics, journalists, and political activists. Ethnic studies programs and organizations like La Raza insinuate an anti-assimilationist bias into the hearts of immigrants and minorities and "shepherd" them into "dependency" and "ethnic ghettos." The sight of thousands of Latino workers marching with Mexican flags and Che Guevara T-shirts on Mayday 2006 was proof that multiculturalism is an impediment to assimilation and a generator of "social calamity." He asserts that within a few short decades "Hispanics" will represent "one-quarter of the American population" and "each time a university president, a politician on the make, or a would-be muckraking journalist chooses the easy path of separatism, he, like the white chauvinists of the past, does his own little part in turning us into Rwanda or Kosovo."[37] For Hansen, multiculturalism facilitates collectivism and strengthens social democratic statism. Catastrophism is the Trojan Horse of South American socialism emanating from the public education system where it has become the ruling ideology.

Tim Donnelly, elected to the House of Representatives as a Tea Party candidate and a founder of the Minutemen Defense Corps of California, sees the threat to America more immediately:

"The facts are incontrovertible that allowing an illegal invasion of the United States will destroy the American Southwest, and very probably wipe out the freedoms we American Christians enjoy, as Muslim Extremists blend in with the so-called 'innocent' illegal aliens, and eventually proselytize them."[38] Conspiracies such as this recall earlier twentieth-century schemes read by the right into Judaism and Catholicism, and parallel the Zionist Occupation Government plots seen by the Aryan Nations and other white supremacist and anti-Semitic groups.

In the United States, the white supremacist movement bases their catastrophic politics around eugenic pseudoscientific arguments for white and Asian intellectual superiority. Their vision parallels the Eurabian terror of Islamic takeover in Europe whereby white American civilization is under threat from a black and Latino majority that will occur sometime around the middle of this century. The effect will be to reduce the national IQ and hasten social collapse and catastrophe. Writers like Byron Roth and Pat Buchanan are tireless promoters of race catastrophe, the semiotics of which is concerned as much with white extinction as with white supremacy. Social programs, affirmative action, and an activist judiciary are the mechanisms by which white civilization is being undermined and hollowed out. According to Buchanan, "our cultural elite allies itself to those out to overthrow the old Christian order—ethnic militants, feminists, atheists—anticipating they will ride the revolution to power. They are succeeding. Our traditional Christian culture has been driven from the temple of our civilization."[39] Buchanan identifies an axis of evil intent on destroying white Christianity. Accordingly, this is really an attack on the white American middle class, a form of cultural Marxism designed to destroy that bedrock of American capitalism. Multiculturalism and a history of democratic expansion and inclusion have eviscerated the ideals of the republic and sullied the constitution under a maelstrom of affirmative action, equal rights, and white victimization.

David Horowitz and others on the right in the United States echo many of Breivik's catastrophe hang-ups around multiculturalism and political correctness. He quotes extensively from the work of British journalist Melanie Phillips, drawing from her assertions that mass immigration to the UK is a Labour Party plot to undermine the political right and to change the nature of British society. Breivik's great catastrophe is the Islamization of Europe and like many rightists he takes Bat Ye'or's book *Eurabia: The Euro-Arab Axis* as his central thesis regarding Muslim immigration to Europe. Ye'or's theory suggests that Muslim immigration into Europe is the consequence of the French elite's wish to consolidate relations with Mediterranean Islamic countries as a counter balance to the power of the United States in Europe. In the early 1970s, these relationships were formalized through intergovernmental arrangements between the EEC (later the EU) and the Arab League culminating in the establishment of the Euro-Arab Dialogue in 1974. The consequences of this pact are the growth of Islamic immigration into Europe and the eventual Islamization of the continent to such an extent that within a generation France will become an Islamic republic. The *Financial Times* characterized the book as akin to *The Protocols of The Elders of Zion* and, while obscure, its theories have gained traction and influence beyond the paranoid fascist margin.[40] Mainstream scholars Daniel Pipes and Niall Ferguson blurb the back cover, Ferguson proposing that "future historians will one day regard her coinage of the term Eurabia as prophetic. Those who wish to live in a free society must be eternally vigilant."[41] The conspiracy is incoherent, not least because it fails to account for the Soviet Union as the key player against which the relationship with the United States was measured in Europe in the 1970s when the plot was allegedly conceived.

Much as the communist threat provided a useful ideological rudder to the militarized economy during the Cold War, so the fantasy of the Islamization of Europe, guided by a liberal elite of politicians, intellectuals, and bureaucrats in conspiratorial league with their Islamic comrades ticks a number of important boxes for the contemporary right. Europe's decline, as outlined by Breivik and echoed throughout the right but in particular by the U.S. neocons, is inevitably the consequence of its aging population, declining birth rates, welfare state, social democracy and liberal morality. The existential threat posed by high Muslim birth rates and unchecked immigration ensures that, as with all useful catastrophic visions, it will happen in our lifetime. The demographic time bomb also plays well with those who promote the idea of a war without end against militant Islam. The Eurabian phantasmagoria finds its context in a "crisis of civilizational morale" where relativism and the abandonment of traditional Christian values sets the stage for Islamic ascendancy in a withering European civilization.[42]

One source for many of these exaggerated fantasies is Fjordman, a well-known far-right Norwegian Internet blogger who writes against Islam, feminism, and multiculturalism.[43] Breivik quotes from him extensively in his "compendium" and is compelled by Fjordman's advocacy of the Eurabia conspiracy. One such story describes how the Inuit people of Greenland are under attack by Muslim immigrants who are attempting to drive them from their homes in Arhus. The Gellerupparken neighborhood has become, for readers of the *Gates of Vienna* blog, a "sharia zone," an area where Islamic courts and cultural groups hold sway and a harbinger of Europe's future where non-Muslims are effectively enslaved.[44] The French Muslim riots and vandalism of the summer of 2006 are another example of "Gaza on Seine," a clear expression of the low intensity warfare waged by Islam in Europe. The riots are directed at the republic itself and are seen as urban guerilla warfare reminiscent of battles waged by Palestinian youth against Israel. "Terror is a way of applying pressure on the European countries to surrender constantly to the Arab representatives' demands. They demand, for example, that Europe always speak out for the Palestinians and against Israel."[45] The French philosopher Alain Finkielkraut described the riots as an "anti-Republican pogrom" in *Ha'aretz*. But it would be a mistake to imagine that this catastrophism is to be found exclusively on the fringe of political discourse in Europe or in America. As already suggested, Bernard Lewis, Ferguson, and Melanie Phillips are as mainstream and prominent as it is possible to be yet all three fully subscribe to the idea that a declining Europe is to be vanquished by a resurgent and imperial Islam.

Support for Israel and U.S./NATO wars against Muslim countries is as widespread among the right in Europe as it is in the United States. Some, like the so-called Antideutsche in Germany, characterize themselves as anti-fascist and regard any criticism of Israel as anti-Semitism while emphasizing their opposition to "Islamofascism."[46] The overwhelmingly secular orientation of the liberation movements that grew in North Africa and the Middle East throughout 2011 have done little to revise their assertion that what we are living through is a clash between the essential cultural backwardness of Muslim peoples and the natural supremacy of the Christian West.[47]

Underlining the Eurabian alarmism of Breivik, Bat Ye'or, Daniel Pipes, and Melanie Phillips is the characterization of Islam as something other than a religion: rather it is an expansionist political ideology that seeks to subsume humanity under the Islamic Umma, a Caliphate of Sharia law extending to the ends of the earth. In this view, Islam is a system of law, morality, and administration without any daylight between the Mosque and the state and any supposed distinction between mainstream and militant or radical Islam is meaningless. The extent to which this analysis has entered the mainstream is indicated by some of the material taught to military officers and civilian agents by the Pentagon. Teaching a course related to the planning and execution of war in 2011, Lieutenant Colonel Matthew Dooley described Muslims thus: "They hate everything you stand for and will never coexist with you, unless you submit." He assumed that the Geneva Conventions had been set aside, opening the possibility "once again of taking war to a civilian population wherever necessary." His conclusion was that Mecca and Medina could be targeted in pursuit of "total war" where Saudi Arabia would be threatened by starvation and "Islam reduced to cult status."[48]

What is remarkable about the European apocalyptic hard right is not just its consistency, but also the versatility with which it has renewed itself since the end of the Cold War. In the 2012 French presidential election, Marine Le Pen, leader of the National Front, came in third with 17.9 percent of the vote. While campaigning, she successfully ditched her party's traditional association with anti-Semitism, Nazism, and holocaust denial, at least in the mainstream French imagination.[49] Instead, she campaigned for "national preference" for French citizens over foreigners for jobs and access to services, opposition to EU integration, and a general critique of the European and French political establishment as a conspiratorial elite. And while there are various fascist associations that endure, like Vichy, she is succeeding in her wish to establish the National Front in the mainstream of French politics. Like Le Pen, the European hard right takes its central platform from these two issues: opposition to further EU integration or outright rejection of the EU, and opposition to further immigration, particularly Muslim immigration. By outlining a dual catastrophe of a centralized dictatorial EU that bakes local and national cultures into a pan European anti identity and the simultaneous overwhelming of European culture by an invading Islamic incubus, the right in Europe has hit on a formula that both appeals to voters and unbalances their electoral enemies.

FN's cyclical appeal to working and lower middle class French voters involves both economic anxieties and resentment toward the political establishment. It is controversial only in that it is attached to a manifesto of anti Muslim and anti immigrant scapegoating: "The progressive Islamisation of our country and the increase in political-religious demands are calling into question the survival of our civilization." Le Pen compared Muslims praying in French streets to the Nazi occupation of France and surveys found that 40 percent of French voters supported the comparisons, more than 50 percent of in the case of conservative voters.[50] The center of French politics has shifted to the right to such a degree that Le Pen can position herself and her party as a populist and "patriotic" antidote to "the totalitarian character of the EU and its desire to remove people's sovereignty."[51] Hostile to the EU and to immigration, the FN has much in common with Austria's Freedom Party, the Italian Northern League, the Danish People's Party, the Swiss People's Party, the True Finns, the Hungarian Jobbik party, the Sweden Democrats, and Geert Wilders's Party for Freedom in the Netherlands. These parties pursue a populist and racist strategy of rhetorical sectarian catastrophe by denouncing the EU and traditional social democratic political establishments as out of touch bureaucratic parasites.[52]

In the 1990s, Dutch far-rightist Pim Fortuyn helped to develop a platform for far-right electoralism that is a model for success in Europe still. Fortuyn was gay and explicitly pro-Israel. He immediately distanced himself from the traditional anti-Semitism and homophobia of European Nazi parties and created the template for a new wave of hard right racists to take on the European mainstream. Fortuyn was assassinated by animal rights activist Volkert van der Graaf while campaigning for the 2002 Dutch general election and his party lost ground to Geert Wilders's Party for Freedom (PVV) without him. Wilders's Islamopanic garnered 15 percent of the Dutch vote in 2010 and earned the PVV a "support agreement" with the right-wing coalition government. Christian Democrat leader Maxime Verhagen assured Dutch radio that "it is possible to come to a stable political cooperation which can have PVV's support."[53] Speaking in Nashville, Wilders warned his audience that "a moderate Islam does not exist and will never exist . . . wake up, Christians of Tennessee. Islam is at your gate."[54] In a documentary film he made in 2008, Wilders is bluntly catastrophic and ends it with a card reading "Islam wants to rule and seeks to destroy our western civilization."[55]

Like the PVV in Holland, the Danish People's Party trades support for a coalition government for influence over immigration policy. In 2005, the party's leader Pia Kjærsgaard told the BBC that if Muslims "want to turn Stockholm, Gothenburg, or Malmö into a Scandinavian Beirut with clan wars, honor killings and gang rapes, let them do it. We can always put a barrier on the Øresund Bridge."[56] Six years later, she got her wish. Denmark abridged the Schengen treaty and reintroduced customs and identification checks on its borders with Germany and Sweden in February 2012.[57] This was the Danish People's Party's price for supporting the center-right government's unpopular pension and welfare cuts. The Danish government insists that the reintroduction of searches and identification checks at the border are to fight cross-border crime, illegal immigration, and drug trafficking while the integration minister Søren Pind advised that the EU needed a frank discussion about the "dark side" of unrestricted travel.[58]

Like the cases of Ye'or and Melanie Phillips, much of the hysteria about the Islamization of Europe is designed not for a European but for an American audience and it is difficult to disentangle it from the broader neoconservative effort to paint Europe as a dangerously liberal redoubt of democratic excess and protosocialist welfarism, eaten away by human rights lawyers, transsexuals, gay adoption, trade unionism, and so on. From a hegemonic business perspective, a supine Europe vulnerable to Islamic domination is the unavoidable result of these same economic and political trends that have undermined Europe's confidence in its own values. From the perspective of those who see the war on terror as an enduring life and death issue for the West, as these writers do, then the catastrophe of Europe's failure and disappearance merely reinforces the idea that the United States is on its own in the battle for good against evil.[59]

Right-Wing Catastrophism and the State

For much of right-wing catastrophism, the state is the site of struggle with and against social movements that promote an expansionist democratic agenda. Cure catastrophists like Christian Identity adherents in the United States believe that the democratic state cannot be disarticulated from historical forces that promote equality and so the state is destined to be overwhelmed via catastrophe if a new order is to be established. Animus toward "government" expressed by the contemporary U.S. right succinctly encapsulates this critique of the state as the vehicle of liberal progress, even if those on the mainstream right who seek to gain control of the state have little intention of actually abolishing it.

The catastrophist right frequently opens political terrain on the far right that the state can exploit. By framing questions like immigration as catastrophic problems, the state is able to respond with harsh and previously off-limits policies. Anti-immigrant sentiment is promoted throughout the European and American center-right, and in both places border "protection" and surveillance are expanding fiefdoms of the security state. Border Patrol is one of the largest law enforcement agencies in the United States. In the EU, or rather outside it, immigrant detention is outsourced to locations beyond the fortress walls, and beyond the news. Political attacks on multiculturalism from the center-right in Europe are in response to a resurgent and increasingly dynamic far right putting these "problems" onto the table.

Since the 1980s, the social democratic and liberal left in Europe and the United States has embraced neoliberalism. It has actively promoted the global expansion of capital and deregulation while attacking the organized working class. But the center-left must also pay homage to ideas of justice and equality, most often directly at odds with their concrete economic policies, or fear a loss of support to competitors further to the left. While extreme right formations have pulled mainstream conservative parties further to the right, the left is also vulnerable to right-wing agitation on questions where the left has traditionally been strong, but is more recently contradicted. Increasingly, these contradictions between economic liberalization and social or economic justice cannot be avoided by center-left parties and this vulnerability provides an opportunity for the right to fragment and disorient the left. In part, for the mainstream right, immigrant scapegoating, along with attacks on multiculturalism and on the economic position of the working class, is an electoral strategy that puts the left and social democratic opposition on the defensive. In periods of economic decline and insecurity, fear generated around these cultural and social issues helps to obscure coherent critiques of economic life. Reproduced as state catastrophism they help to reinforce this fragmentation of the left. Fear is the bedfellow of right-wing catastrophism and it is expertly manipulated by the state.

A right-wing catastrophic vision of the state extends back at least to Thomas Hobbes. Hobbes's *Leviathan* posits civil war as the state of nature. To tame and achieve dominion over that hellish prospect, Hobbes envisions a "commonwealth" of men held together by contract and the dictatorial power of the sovereign who will "punish with corporal or pecuniary punishment or ignominy" those who fail to abide by the rules.[60] It stands in contrast to the more optimistic prescriptions of Montesquieu and Rousseau that articulated the rule of law, and its implicit countervailing foci of power, over the rule of men. The Enlightenment crystallized a set of ideas that had been forming in Europe since Martin Luther. It began as a break with the dictatorial rule of the Pope and the idea of individual political liberty, and the individual that the Enlightenment presented remains a central doctrine of democratic political thought.[61]

Hobbes emphasized (and this is why he remains a touchstone of authoritarian ideology) that violence is at the heart of politics. There is only a choice between the ordered monopoly on violence that codifies a set of sociopolitical arrangements, the state, or an inchoate and tempestuous violence that sets each against all. Hobbes is the founder of a right-wing tradition that has as its core assumption a notion of human nature as predatory and selfish and this tradition gets made and remade against Rousseau and certain versions of Enlightenment thought right up to today. Hobbes is not the first catastrophist, but he was perhaps the best known and, at the dawn of capitalism, identified catastrophe as the disease for which the state is a cure. Conservatives hold him in special regard and there are few liberals who are not Hobbesians at heart either: Hillary Clinton, Tony Blair, and Barack Obama would all agree that the state of nature is war.

If state power is an apparatus, then catastrophism is employed to expand it. States recognize that catastrophe must be averted by whatever means are available, and they understand that catastrophe aversion is a powerful political multiplier. FDR used it when he sold the New Deal, and Hitler proclaimed it when he invaded Poland. In the section "Creating Tomorrow's Dominant Force," the Project for the New American Century indicated that their transformative project to remake an American world was likely to be a long one "absent some catastrophic and catalyzing event—like a new Pearl Harbor."[62] Conspiracists mistake that to indicate a scripting of the attacks of September 11, 2001, but it can be more accurately read as the simple recognition that catastrophe or the threat of catastrophe makes radical change possible. Goering pointed out that when it comes to persuading people of the need to go to war, "all you have to do is tell them they are being attacked, and denounce the pacifists."[63] Catastrophe can then be anything that makes war look like the better option.

State catastrophism often indicates an exchange of political and social freedoms for relief from fear.[64] The aftermath of September 11 confirms the idea that the promotion and management of fear is the foremost technique for those who wish to exercise control over events. The saturation of American culture with reminiscences of World War II in the years following 9/11 was an extroverted yearning for a "popular" emergency and simultaneously a bulwark for an unpopular one. Confronted with a threat or the imaginary landscape of threats, the state, like the sovereign, allows itself room for exceptional action and response. In the arena of hypothetical disasters, the state does not resort to a judicial standard of proof, instead "urgent conjecture must sometimes take the place of proof" and the greater the supposed threat the greater the onus on the skeptic to prove a negative, that there is no threat.[65]

The detonation of the first Soviet atomic bomb in 1949 shocked Western intelligence services and disrupted U.S. plans to contain the Soviets to the territory they controlled at the end of World War II. The National Security Council warned President Truman that, "The United States . . . is the principal enemy whose integrity and vitality must be subverted or destroyed by one means or another if the Kremlin is to achieve its fundamental design."[66] The NSC assumed that war with the Soviets was inevitable and, when fighting broke out in Korea in 1950, its catastrophist interpretation of Soviet foreign policy became the governing American orthodoxy. Truman declared a state of emergency and four troop divisions were dispatched to Germany: "The issues that face us are momentous, involving the fulfillment or destruction not only of this Republic but of civilization itself."[67] Within two years, the United States would relocate massive air, ground, and naval forces to Europe and, guided by NSC doctrine, embark on a long-term strategy to intensify the global military and political struggle with communism under the banner of the Cold War and atomic Armageddon. "The present world situation, however, is one which militates against successful negotiations with the Kremlin, for the terms of agreements on important pending issues would reflect present realities and would therefore be unacceptable, if not disastrous, to the United States and the rest of the free world."[68] George Kennan, then a senior Soviet analyst in the State Department, demurred from the NSC line arguing that Stalin had no taste for territory where he could not have direct political or military control, but it was too late. The ship had sailed with Truman to the Cold War.[69]

Asked after the invasion of Iraq in 2003 to outline the difference between an assumption that Iraq had WMDs and the hypothesis that Saddam might "move to acquire those weapons," President Bush responded: "So what's the difference?" His refusal to recognize the distinction between the possibility of a fact and the possibility of a desire to establish facts may have seemed novel, but it is consistent in the history of exceptions to the supposedly normal course of state action. Writing about the Weimar constitution in 1921, Carl Schmitt observed that the ability of the president to declare a "state of emergency" implicitly recognized that dictatorial power was a feature of the state. He later distinguished an enemy of the state as "in a specially intense way, existentially something different and alien, so that in the extreme case conflicts with him are possible."[70] States routinely deploy emergency powers to deal with the vagaries of nature and disaster but these instances are generally temporary and localized. It is in the case of war that emergency power is most spectacularly invoked. In Giorgio Agamben's interpretation, the kind of violence Schmitt valorizes transforms the state into an "apparatus of death" and "the state of emergency defines a regime of the law within which the norm is valid but cannot be applied (since it has no force), and where acts that do not have the value of law acquire the force of law."[71] Schmitt argued that it is impossible to predict the nature of threats or the conditions of any emergency so it is impossible to prescribe any legal form or limit to sovereign action. Post 9/11, Schmitt's key insight plays on in the procedural gymnastics over torture, assassination, kidnapping, drone murder, domestic spying, extra judicial internment, and the limits of executive power. Until the twenty-first century, one could still assume that the United States was constitutionally distinct from any of Hobbes's prescriptions. But Dick Cheney's contention that presidential power and what the president does is the same thing echoes Schmitt (and Hobbes) from beyond the grave. Schmitt's assertion that it is "precisely the exception that makes relevant the subject of sovereignty" recognizes that the exception becomes the rule and for him confirms that fascism is not incompatible with democracy.

Henry Paulson's TARP fund for the financial sector following the collapse of Lehman Brothers in 2008 was a catastrophe of another sort. Announcing the program, Paulson stated that the initial $700 billion would be used to relieve banks of worthless mortgage-backed securities, but the bill never specified how these jubilee funds would be spent. The threat that Paulson identified was the insolvency of many banks as a result of the collapse in value of their mortgage holdings, and the consequent "clogging up" of lending and money markets. Within weeks though, banks were using the money to consolidate and buy other banks rather than sanitizing their capital base. Paulson's threat to "the financial security of all Americans—their retirement savings, their home values, their ability to borrow" was not averted.[72] But neither did the economy collapse. As much as the threats may or may not be real, the outcomes may not be as intended or the intended outcomes may not be as stated. In 2008, the invocation of imminent catastrophe allowed for a departure from normal procedures necessitating secrecy, speed, and huge quantities of money. Brad Sherman of the House of Representatives from California was warned that "the market would drop two or three thousand points the first day, another couple thousand the second day, and a few members were even told that there would be martial law in America if we voted no."[73]

* * *

Catastrophism has a long history on the right and both the state and the organized far right understand it and wield it skillfully to achieve political and propaganda goals, as this chapter has sought to show. Disease catastrophists view the achievements of the left as suicidal threats to traditional order and this view is universal on the right.[74] Cure catastrophists believe that violent conflict will resolve and defeat these threats and some among that group, like Breivik and Timothy McVeigh, desire to quicken its arrival.[75] In the contemporary period, marked by persistent economic and environmental crises, catastrophes are more visible and their invocation is even more common. For the left, as outlined elsewhere in this book, this presents serious problems in trying to form a political understanding that is useful for organizing and for social movements. But for the right, catastrophism is not counterproductive. As we have seen, it allows the right to influence and even dominate many political questions and at the state level real victories are being achieved.

Catastrophism is a less ambivalent strategy for the right than for its adversaries on the left. From a rhetorical standpoint, catastrophism is a win/win for the right as there is no accountability for false prophecy. On the one hand, it rallies the troops and creates a sense of urgency. On the other hand, though, fear and paranoia serve a rightist political predisposition more than a left or liberal one. Authoritarian politics benefits more than left politics from fear. Twenty-first-century capitalism is characterized by a high degree of insecurity for all workers, both middle and working class, and fear and vulnerability constitute a growing part of the social landscape. The right can profit from exploiting these conditions, and in light of their achievements over the last generation they will continue to do so. Ironically, it is the collapse of the left that has offered up the space for them to do it, so the weeds are growing in the beloved ruins of the Keynesian state. The right has built a popular opposition to the welfare state and to income redistribution by shaping resentment against minority groups. Catastrophism for the right is the fight against equality and for war, hierarchy, and state violence. In a thoroughgoing way, right-wing catastrophe manages to materialize Samuel

Huntington's prediction that "the fundamental source of conflict in this new world will not be primarily ideological or primarily economic. The great divisions among humankind and the dominating source of conflict will be cultural."[76]

Catastrophism is one way to shift the focus from the essential questions of public policy, democracy, equality, access to education and health, environment, etc. and onto abstractions about civilization, culture, and threats to the prevailing social order that promise instability and worse. It is ironic that the contemporary right has found in identity a politics to sustain itself, much like it charges the left with having done. But in a period of declining incomes, chaotic public finances, and persistently high unemployment, the promises and predictions of a "catastrophically convulsed America that descends into a Balkanized ruin and social collapse" seeps into the right-wing zeitgeist and finds scapegoats by successfully fragmenting an already fragile and divided political landscape. Which, of course, is the point.[77]

CHAPTER FOUR

Land of the Living Dead: Capitalism and the Catastrophes of Everyday Life

David McNally

That things are "status quo" is the catastrophe.
—Walter Benjamin[1]

CATASTROPHIST ANXIETIES HAVE REMARKABLE REACH ACROSS THE CULtural space of late capitalism. More than merely the nightmare scenarios of apocalyptic preachers and prognosticators, predictions of impending doom are also found in the writings of thoughtful social commentators and critics.[2] But nowhere do catastrophe and apocalypse loom larger than in film and fiction, particularly in the horror genre, where zombies and vampires fill theaters and fly off bookstore shelves. Zombies are a particular rage these days, having made so indelible a mark on mass culture during the global economic crisis of 2008–2009 that *Time* magazine declared them "the official monster of the recession."[3]

The cultural omnipresence of zombies and other monsters offers a clue as to the mysteries of everyday life in capitalist society. For nearly two hundred years, specific imageries of horror—dissection, body-snatching, dismemberment, and blood-sucking—have haunted the popular imagination, hinting at a profound sense of corporeal vulnerability intrinsic to modern life. And today, in the context of a global economic slump, persistent wars, and worsening environmental crises, many of these imageries have taken on an apocalyptic hue.

But it is the mundane rather than the apocalyptic figures of monstrosity that most concern me here. In many ways, the world's most obvious horrors—genocides, mass displacements, famines, wars, and ecological calamities—are easily identified, even if the dominant ideology works overtime to distort and mystify them, refusing to disclose their connection to capitalist structures of power. But beyond these overwhelming public atrocities, our culture also seethes with anxiety about the largely prosaic and unacknowledged catastrophes of everyday life. And this is where the proliferation of zombies and vampires across the cultural landscape can become significant: as registers in which these banal horrors are recorded, albeit in partial and distorted form. As much as these monsters are frequently absorbed by the ritualized formulas of the culture industry, they contain an unabsorbed "surplus" of meaning that speaks to deeper truths. For lurking within these commercialized images is the eerie sense that monstrosity lies very close to home indeed, in the most ordinary practices of everyday life—in other words, that there are strange and chilling things happening right around us, *to us*. The very everydayness of grotesque images is a warning that ominous forces are not just "out there," in regions of the strange and the horrifying, but *in here*, in the very spaces through which we move, invading our bodies and minds.

"The monster terrifies because 'it' represent the terrifying fate of our own bodies," one historian astutely observes.[4] This existential reality—the disquieting fate of our own bodies in capitalist society—helps explain the mass production of images of zombies that dismember us and eat our flesh, of vampires that suck our blood, or body-snatchers that possess our helpless anatomies. In what follows, I try to make sense of these catastrophic imaginings of everyday corporeal vulnerability that drive so much cultural production today. In examining two of the most enduring monsters of our epoch—Frankenstein's Creature and the zombie—I hope to decipher what it is about capitalism that makes body panics an epidemic feature of everyday life. I then trace two contemporary variants of the zombie—the cannibalistic consumer and the living-dead laborer—to see what, taken together, they tell us about our current historical moment. Finally, I explore the rebellious underside to our obsession with monsters in order to illuminate a utopian impulse that hints at ways out of late capitalism's nightmarish landscape of ruin.

Frankenstein, the Corpse Economy, and the Living Dead

"Horror stories," observes anthropologist David Graeber, "always seem to reflect some aspect of the tellers' own social lives."[5] This is particularly true of arguably the most famous monster in the history of capitalism, Victor Frankenstein's Creature. Since its literary emergence in 1818 in Mary Shelley's acclaimed novel, the Creature has become an enduring cultural referent, recirculated in every available medium, from theater to film, from novels to comic books and television cartoons.[6] In pondering the popularity of this figure, however, commentators have rarely probed the unique mixing of the horrors of dissection and grave-robbing that animate the story—and which clearly accounts for its profound resonance in the nineteenth century and beyond.[7]

Interestingly, Mary Shelley is explicit about these themes. Early on in her novel, Victor Frankenstein informs us that he is both an anatomist and a grave-robber. "I became acquainted with the science of anatomy," he explains, and spent "days and nights in vaults and charnel houses."[8] Combining body parts stolen from corpses with bits of dissected animals, he cobbles together his monstrous creation:

> Who shall conceive the horrors of my secret toil, as I dabbled among the unhallowed damps of the grave, or tortured the living animal to animate the lifeless clay? . . . I collected bones from charnel houses; and disturbed, with profane fingers, the tremendous secrets of the human frame. . . . The dissecting room and the slaughter-house furnished many of my materials.[9]

Readers today often miss the social significance of these themes, but early nineteenth-century audiences did not, for during the rise of capitalism, dissection had become an instrument for punishment of the poor, and grave-robbing a marker of the ill-treatment of plebeian bodies. As Peter Linebaugh has brilliantly shown, saving the bodies of the condemned from dissection was a central feature of a working class culture of solidarity that challenged the law and the authorities in eighteenth-century London.[10] To this end, crowds regularly rioted at the gallows, engaging in pitched battles, sometimes lasting for hours, all in order to keep the bodies of the hanged out of the hands of "anatomists" intent on dissecting them.

The riot at the gallows was one of the most distinctive features of social life in eighteenth-century London, the urban center of emergent capitalism. Its presence dramatically signaled a deeply felt anxiety about the integrity of the laboring body. And the poor had good reason to be worried about their bodily well-being.

As capitalist relations became entrenched, so in ruling class circles did the idea that the poor had no right to life beyond what they could secure in the market. Traditional forms of relief for the poor were scaled back, and whipping, branding, and confinement for "crimes" such as begging became part of the punitive armory of ruling class power. So did dissection. In 1694, the London town council decreed that abandoned bodies of the poor—found dead in the street or unclaimed after violent deaths—could be provided to the anatomists. Still, as the practice of anatomy boomed and medical education increasingly used dissection, the supply of corpses failed to keep pace. The result was twofold: first, a steady rise in the price of corpses, which more than tripled in the twenty years after 1720; and, second, growth in the practices by which they were illicitly procured—grave-robbing, murder, and the purchase (from relatives and friends) of bodies of the condemned on hanging days outside London's Newgate Prison. The class dimension of all this was blatant: it was the bodies of the poor, and the poor alone, that were up for grabs.

By the 1720s, corpse-stealing had become a full-time profession practiced by "resurrectionists," who could make a comfortable living at the trade. And, as the market increased, so did evidence of murder, particularly of street youth, in order to sell their cadavers for dissection. The result was a *corpse economy* in which human bodies, increasingly commodified in life, became literal commodities in death. To commodify a living being is of course also to reify it, to treat it as an inanimate object. In the case of the corpse economy, so extreme was this reification that a cadaver intended for sale was dubbed a "Thing." And as befits a capitalist market, the commodity-corpse was subjected to pricing policies as subtle as those applied to any good. As historian Ruth Richardson observes,

> Corpses were bought and sold, they were touted, priced, haggled over, negotiated for, discussed in terms of supply and demand, delivered, imported, exported, transported. Human bodies were compressed into boxes, packed in sawdust, packed in hay, trussed up in sacks, roped up like hams, sewn in canvas, packed in cases, casks, barrels, crates and hampers, salted, pickled, or injected with preservative. . . . Human bodies were dismembered and sold in pieces, or measured and sold by the inch.[11]

Popular preoccupations with grave-robbing and dissection owed much to the ways in which this treatment of working class bodies in death simply mirrored their daily life experiences. Crucial here is the fact that capitalism requires that everything, particularly people's embodied working energies (or labor power), can be bought and sold. Even where workers are not sold once and for all, as are slaves, they are nonetheless expected to package up discrete parts of their lives and their life-energies (to dissect them as it were) and offer them for sale to an employer. Workers must thus treat parts of their humanity as dead things, as mere commodities, available to the highest bidder. As Marx put it,

> The exercise of labour-power, labour, is the worker's own life-activity, the manifestation of his own life. And this life-activity he sells to another person in order to secure the necessary means of subsistence. Thus his life-activity is for him only a means to enable him to exist. He works in order to live. He does not even reckon labour as part of his life; it is rather a sacrifice of his life. It is a commodity which he has made over to another . . . life begins for him where this activity ceases, at table, in the public house, in bed.[12]

But if life begins outside work, then this means that work, the period in which our energies are sold to someone else, is a sort of death, an absence of life. During our working hours, we experience "dead time," a time of nonlife, a sort of *living death.*[13] This is why the corpse economy became such a powerful symbolic register for working people's anger about the market system. Just as the human corpse was becoming a new kind of commodity throughout the eighteenth century, so were the living bodies of the poor. So, as capitalism reduced the human body to a (generally mindless) assemblage of parts to be harnessed for production, workers were often referred to as "hands"—sometimes with qualifiers, as in dockhands, farmhands, hired-hands, and so on—in short, as body parts.

This is why anatomists symbolized everything the poor loathed about the new market economy. Body-snatching, dissection, and the trade in corpses were proof that the monstrosities of the market respected no limits; they demonstrated that the market economy happily embraced what one trade unionist called the "odious and disgusting traffic in human flesh."[14] The corpse-economy thus became emblematic of all that was objectionable about emergent capitalism, of its demonic drive to exploit human life and labor, of its propensity to humiliate and demean in both life and death.[15] In wresting a corpse from the surgeons, the crowd thus struck a blow—both symbolic and real—against commodification and for the integrity of the proletarian body, if only in death. In burying it intact, they claimed a moral victory over the dismembering powers of capital.

By aligning Victor Frankenstein with the surgeons, anatomists, and grave-robbers, Mary Shelley touched on some of the deepest fears of the laboring poor during the rise of capitalism. At the same time, her novel was also meant to provoke the fears of the rich and powerful. If the anxieties of the poor were prodded by images of grave-robbing and dissection, the figure of the massive and insurrectionary proletarian troubled Britain's rulers. In her construction of the Creature, Shelley played on both sets of anxieties. In Victor Frankenstein's assembly of the monster, for instance, Shelley imaginatively reconstructed the process by which the working class was created: first dissected (separated from the land and their communities), then reassembled as a frightening collective entity, that grotesque conglomeration known as the proletarian mob. "Like the proletariat," notes Moretti, "the monster is denied a name and an individuality. . . . Like the proletariat, he is a *collective* and an *artificial* creature."[16] In so depicting the monster, Shelley cautioned the ruling class that they had conjured up a gigantic being with immense destructive power. And while it may have been an artificial creature, in Shelley's telling it was also an intelligent and articulate one.

Contrary to most film versions in which the Creature is typically a mute brute, Shelley portrays him as an intelligent being with linguistic capacities.[17] This decision highlights the monster's humanity and radically demarcates him from the typical image of the zombie. In fact, in one of the most celebrated film versions, the most famous Hollywood actor to play the Creature, Boris Karloff, deliberately *zombified* him. Karloff gave the Creature the shuffling gait we now associate with zombies and vigorously opposed letting the monster speak. "If he spoke, he would seem more human," he argued.[18] Yet, this was precisely Shelley's point: the working class was both grotesque (monstrously oppressed and impoverished, and seething with anger) and intelligent. As frightening as the Creature might be, his capacity for speech is a fundamental marker of his humanity, of the fact that proletarians are conscious and articulate members of humankind—i.e., *not* zombies. Conferring language on the Creature is essential to the central hinge of the novel, a lengthy speech in which the monster narrates his life experience and stakes his claim for justice. In subsequently stripping the Creature of speech and intelligence, and in giving him a slow, ambling gait, modern versions made him something closer to a zombie, that other monster who tells us so much about life in capitalist society.

Zombies and the Living-Dead World of Work

More than anything else, the earliest modern images of the zombie are tied to figures of mindless labor. The cultural roots of the *nzambi* are to be found in West African belief systems (specifically associated with the lower Congo), which held that the dead could return to visit their families, bringing either assistance or harm.[19] During the era of the slave trade, the image seems to have been adapted to the horrors that tore people from their communities and their kin, stripping them of their "selves," and reducing them to mere laboring flesh for sale. In Dahomean culture, according to one anthropologist, zombies were portrayed as creatures put into a death-like state by sorcerers only to be dug up later from their graves and sold into slavery.[20] But it was in Haiti, where by 1789 half a million slaves toiled on French plantations in conditions similar to industrial labor, that the zombie was definitively transmuted into a figure of extreme reification—a living laborer capable of drudgery on behalf of others, but entirely lacking in memory, self-consciousness, identity, and agency, the very qualities we associate with personhood. [21] Moreover, zombie legends acquired a special timbre during the period of American occupation of Haiti (1915–1934), when U.S. marines, dominating by violence and terror, deployed forced labor to build roads and other infrastructure.[22] In modern Haiti, zombies thus acquired their unique meaning as the animated dead, mere flesh and bones toiling on behalf of others.[23] Significantly, it was this version that passed into American culture during the Great Depression as a result of a William Seabrook's highly influential book, *The Magic Island* (1929), written during the American occupation, after the author spent a year with a Haitian family that allegedly initiated him into voodoo.

Though overflowing with ethnocentric stereotypes, Seabrook's book nonetheless offered a highly poetic account of zombies that reverberated across Depression-era America and formed the basis for the creatures' earliest filmic representations. In a chapter entitled "Dead Men Walking in the Cane Fields," Seabrook recounts a friend's putative response to a question about "zombie superstition" in Haiti as follows: "Alas these things—and other evil practices connected with the dead—exist. . . . At this very moment, in the moonlight, there are *zombies* working on this island. . . . If you will ride with me tomorrow night, yes, I will show you dead men working in the cane fields." Especially interesting is the friend's further allegation that zombies work in the fields of the Haitian-American Sugar Company, a firm whose main plant is described as "an immense factory plant, dominated by a huge chimney, with clanging machinery, steam whistles, freight cars."[24]

Finally, when he is taken to witness the creatures for himself, Seabrook writes that he observed

> three supposed zombies, who continued dumbly at work . . . there was something about them unnatural and strange. They were plodding like brutes, automatons. Without stooping down, I could not fully see their faces, which were bent expressionless over their work. . . . The eyes were the worst. . . . They were in truth like the eyes of a dead man, not blind, not staring, unfocused, unseeing. The whole face, for that matter, was bad enough. It was vacant, as if there was nothing behind it. It seemed not only expressionless, but incapable of expression.[25]

Whatever we make of this account, which clearly mobilized key Haitian tropes about the living-dead, Seabrook's tale became the point of reference for the earliest depictions of zombies in American literature and film. Moreover, the central features of his rendition converge with those found in another influential portrayal, that of Alfred Métraux who wrote in his book, *Le Vaudou Haitien* (1957): "The zombie remains in that grey area separating life and death. He moves, eats, hears, even speaks, but has no memory and is not aware of his condition. The zombie is a beast of burden exploited mercilessly by his master who forces him to toil in his fields, crushes him with work, and whips him at the slightest of pretexts."[26]

This view of zombies—as mindless laborers—entered the American culture industry in the 1930s and 1940s. Furthermore, this image carried a latent but powerful social criticism: the idea that in capitalist society the majority become nothing but bearers of undifferentiated life energies, dispensed in units of abstract time. The *raison d'être* of zombies is the labor they perform. They themselves are meaningless, interchangeable beasts of burden, who exist as mere bodies for the expenditure of labor-time. For the zombie, as for the wage-laborer, "Time is everything, man is nothing; he is at the most, time's carcase," as Marx memorably put it.[27] Here is where we see the point of intersection between the experiences of slavery and wage-labor, manifestly distinct though they are. If, as Orlando Patterson powerfully argues, the slave is socially dead, the wage-laborer is periodically and repeatedly so.[28] If the slave is permanently zombified, the wage-labor is subjected to a zombie repetition compulsion.[29] This is why the zombie image is capable of resonating across social contexts involving both slavery and wage-labor. For in both contexts, people identify with the zombie as a "mythic symbol of alienation . . . of the dispossession of the self through the reduction of the self to a mere source of labour."[30]

But this idea of the zombie as a living-dead laborer was displaced in American cultural production in the late 1960s by that of the ghoulish consumer. While this cultural shift is intriguing and occasionally sharply satirical, it nonetheless dulled the most critical edge of the image, although the latter persists in zombie tales proliferating throughout sub-Saharan Africa in the neoliberal era.[31] To put it plainly, if Hollywood's zombies today have largely become mindless consumers, in Africa they remain mindless workers. Profound truths about the world of late capitalism lie at the junction of the two images.

Crazed Consumers and Lifeless Laborers: Images of the Zombie in the Neoliberal Era

A significant transformation took place in the Hollywood image of the zombie during the rebellious period of black, working class, feminist, and antiwar protest of the late 1960s and early 1970s. Crucial here was the displacement of the image of the zombie-laborer by that of the manic consumer. A key symbolic shift involved the attribution of flesh-eating to zombies, a move that was consecrated in George A. Romero's *Night of the Living Dead* (1968), despite the fact that Romero considered his flesh-eaters to be "ghouls," not zombies. Indeed, the original title of the film was *Night of the Flesh Eaters*, a label that was dropped because of its proximity to that of another movie. But with the addition of the living dead to the title, zombies became newly identified with cannibalism.[32] While zombies remained dull and plodding, they now appeared as manic consumers (of human flesh) rather than mindless laborers. Of course, for such an identification to receive wide play, broader cultural shifts had also to be at work, in particular an association of hyperconsumption with a Western capitalism that in bringing cars, televisions, and household appliances to millions also bred violence and impoverished the actual quality of life.

In this context, *Night of the Living Dead* carried a significant critical charge. As film critic Robin Wood has suggested, Romero's zombie-ghouls represent capitalists, and "cannibalism represents the ultimate in possessiveness, hence the logical end of human relations under capitalism." Wood sees the victims of the flesh-eaters as all those who are outcast and oppressed in American society, including feminists, gays and lesbians, and Civil Rights activists.[33] Other commentators astutely observed that the film also operated as a biting critique of the American state's war in Vietnam, portraying America's armed forces as flesh-eating monsters. A critic with the *Village Voice*, for instance, noted that the movie "was not set in Transylvania, but Pennsylvania," a new region of middle-American horror, and that "the zombie carnage seemed a grotesque echo of the conflict then raging in Vietnam." In addition, the most prominent heroic character in the film, its sole African-American man, "survives the zombies only to be killed by a redneck posse," as if mirroring racist violence against Civil Rights and Black Power activists.[34]

But as the New Left of the 1960s and early 1970s went into decline, along with the mass social movements that had sustained it, it became more difficult for such zombie imagery to operate as potent social criticism. Of course, cannibalism could still function as a stinging metaphor for a First World capitalism choking on its own excesses. And there is certainly a nice satire along these lines at work in Romero's 1978 zombie film, *Dawn of the Dead*, the bulk of which takes place in a shopping mall to which the creatures are obsessively drawn. Yet, the association of late capitalism with cannibalism and manic consumption all too easily became cliché, something to be enjoyed over movies and popcorn without provoking wider critical analysis. After all, it is not consumption that drives a capitalist economy. Instead, the secret of the system is profit-generating labor. While crazed flesh-eating could in the right context serve as a metaphor for an insane war machine, in the absence of mass antiwar and social movements, it quickly lost that edge. Detached from the figure of the zombie-laborer as a beast of burden controlled by sinister forces, Hollywood's zombie films lost much of their critical charge. At the same time, a new genre of zombie images, emanating from sub-Saharan Africa, remobilized key ingredients of Haitian representations of zombie-laborers, and in so doing reinstated their critical possibilities.

Thus, if it is true that "The only modern myth is the myth of zombies—mortified schizos, good for work, brought back to reason," it is in sub-Saharan Africa that this truth has been most powerfully rendered.[35] Intriguingly, these African zombie tales reverse Hollywood's cultural shift, moving from stories of sorcery based on cannibalism to a newer type based on the exploitation of labor. This transition occurred decisively during the era of capitalist globalization, beginning in the 1970s, when a new sorcery appropriate to capitalist relations in urban settings massively proliferated throughout oral culture, pulp fiction, film, and video.[36]

The return of the zombie-laborer fits with decisive changes that accompanied the neoliberal period of capitalist globalization, when commodification and monetized relations more fully penetrated most African societies. In a context of mounting sovereign debts, Structural Adjustment programs, cuts to subsidies for the poor, privatization, and the opening of national economies more fully to global capital, everyday life became increasingly subordinated to the imperatives of the market. One cultural expression of these changes was a new genre of monster tales.

In many older African discourses of sorcery (or "witchcraft"[37]), the typical occult individual was a crazed, cannibalistic consumer. Such folktales presupposed a zero-sum local economy in which one person's loss was another's gain: what the witch ate or accumulated could only come at the expense of a neighbor. Such imaginings are widespread in noncapitalist societies, but they suffer an imaginative and explanatory failure when people encounter the ever-increasing circuits of global capitalist accumulation. The apparently infinite capacity of capitalism to expand is simply not explicable in the terms of this social imaginary; it is inconceivable that the vast wealth of multinational firms and *nouveaux riches* elites could come from theft within village communities. Thus, as capitalist globalization has imposed market logics ever more directly on Africa's peoples (with millions pushed off the land under the impact of structural adjustment and displacement, swelling urban slums dominated by all the depredations of peripheral capitalism), older bewitched economies have bumped up against their explanatory limits.[38] As a result, the Ihanzu of north-central Tanzania, for example, now speak of something that exceeds traditional witchcraft, a new mode of occult accumulation driven by "business witches" who do not need to devour in order to acquire.[39] Magically transcending the limits of the zero-sum game, this new sorcery is capable of infinite wealth creation, imagined in terms of great mountains of commodities and money. For the Temne of Sierra Leone, a new "economic witchcraft" pivots on invisible transactions that move money from people's pockets to those of witches, or store it in an invisible "witch-city," a place of skyscrapers, luxury cars, airports, VCRs, and street vendors who sell human meat on a stick.[40]

Similarly novel witchcrafts have emerged in Cameroon, where new occult economies revolve around "a witchcraft of labour."[41] Among the Bakweri, the new breed of witches may still kill their victims, but rather than eat them, they convert them into zombie-laborers. Similar ideas are found in a new urban magic among the Douala, wherein people are said to be sold rather than eaten, with the profits flowing directly to the witches' bank accounts.[42] Legends also abound in Cameroon, South Africa and Tanzania of zombie-laborers toiling on invisible plantations in an obscure nighttime economy. These include tales of part-time zombies, captured during their sleeping hours, only to wake up exhausted after their nocturnal labors.[43]

There is no doubt that all such imaginings are multivalent, weaving together diverse strands of human experience—histories of race, gender, class, and kinship; memories of slavery, colonialism, and war; experiences of marketized and monetized social relations; the savage consequences of Structural Adjustment Programs; the corruption of postcolonial elites; the devastation wreaked by an AIDS pandemic—into coherent local discourses. But there can be little doubt that they also speak eloquently to experiences of intensified commodification and new patterns of accumulation that accompany neoliberal globalization. Indeed, as one commentator observes, despite many differences of imagery and nuance, all these new discourses of the occult share the assumption "that witches no longer see their fellow men as meat to be eaten . . . but rather as laborers to be exploited."[44]

We witness here a shift from use value to exchange value: rather than employing sorcery for immediate consumption, modern economic witches instead harness the productive capacities of victims for purposes of accumulation. In all these cases, bodies are not absorbed *into* the other, but transformed into *extensions* of the other—into forces of production (zombie-laborers) or commodities for exchange. The expansive dynamic of capitalism intrudes here into the very grammar of sorcery. No longer does demonic greed revolve around the simple appropriation and consumption of finite social wealth; no longer are witches simply cannibals. With the increasingly ruthless subordination of African economies to the logic of neoliberalism, via the combination of Structural Adjustment programs and world market pressures, witchcraft in urban sub-Saharan Africa has now entered the limitless circuits of global capital. And this has brought about a heightened emphasis on zombies as living-dead laborers who make the magic of accumulation possible.

It is not surprising that the African subcontinent is the space in which such zombie tales have come to life. After all, no region of the global economy has been as catastrophically savaged by neoliberalism. Perhaps the most staggering statistic in this regard comes from the World Bank itself: between 1987 and 2000 per capita incomes in sub-Saharan Africa contracted by fully 25 percent.[45] Yet even that statistic is too rosy: currency devaluations required by international financial institutions have sent the costs of basic goods like bread and heating oil soaring, while the speculation induced by financial liberalization has driven rents and housing costs astronomically higher.

Meanwhile, collapsing prices for agricultural products force millions to abandon the land and head for already overcrowded cities lacking adequate housing, sanitation, and running water. Countries like the Ivory Coast, Nigeria, and Congo-Brazzaville, once classified as "medium income countries," have experienced a horrifying regression, as poverty rates rise and life expectancy plummets. All told, indices of human development are regressing in fourteen countries on the African subcontinent,[46] undoubtedly worsening in light of the global economic slump that broke out in 2008.[47]

In the face of this genuine human calamity, tragically exacerbated by debilitating payments on foreign debts, popular culture has generated tropes for understanding these processes of commodification, exploitation, and impoverishment.[48] Throughout the African subcontinent, the figure of the zombie-laborer has come to depict the dirty secret of late capitalism: that rather than a high-tech world of frictionless circuits of accumulation, capitalism continues to subsist on hidden sites of sweated labor. By focusing on labor rather than just consumption, these monster tales retain a critical edge. Across these stories, real bodies are implicated: they perform unseen labor; they are possessed by evil spirits that turn them into money machines; they are dissected for marketable parts. For all their involvement with witches and spirits, these tales are driven by a materialist impulse to search out the unseen sites where laboring bodies are exploited and at risk. And in seeking out those bodies, and the ways they are incorporated into concealed processes of exploitation, these African discourses of witchcraft describe dangerous logics of accumulation—nowhere more life-threatening than in sub-Saharan Africa itself.

In light of the age of austerity that now grips the Global North and the declining living standards and mounting hardship it entails, this is a propitious moment for a dialectical encounter between the two contemporary zombie figures we have been exploring. In the picture of the maniacally insatiable flesh-eater, we find the capitalist-zombie, driven to relentlessly consume human beings. Meanwhile, in the image of the zombie-laborer we encounter the reality of the global collective worker reduced to a beast of burden who keeps the machinery of accumulation ticking. These zombie figures are in fact genuine doubles, creatures whose existence is inextricably bound up with the other. Taken together, they define the zombie-economy of late capitalism, an out-of-control, flesh-eating machinery of manic accumulation and exploitation that has become an end in itself, driven ever onward toward what we all suspect will be a nightmare ending, a desolated postapocalyptic *Blade Runner* world of rotting factories, environmental destruction, marauding armies, and dying cities.

But the clash of the manic flesh-eater and the laboring drone also hints at another startling zombie capacity: rebellion. For the image of zombie revolt, of polite society being turned upside down by legions of disheveled, marauding monsters, rejoins the most subversive impulses of the original story of Frankenstein and his Creature. After all, the monster's capacity to resist, indeed to overturn the social order, constituted its threat to bourgeois authority. And in that figure of rebellion we find the utopian element of late capitalist catastrophism—the faint but persistent image of zombies on the march, awakening to consciousness, and turning the world upside down.

The People under the Stairs and the Everyday Work of Resistance

Of course, little is served by treating apocalypse itself as a world-changing event. End-of-the-world panic is most likely to be debilitating rather than energizing, as other authors in this volume point out. The truly subversive image of zombie revolt in fact returns us to the everyday—to the idea that revolution grows out of ordinary, prosaic acts of organizing and resistance whose coalescence produces a mass upheaval. However extraordinary a popular uprising may be, it is nonetheless a product of decidedly mundane activities—strikes, demonstrations, meetings, speeches, leaflets, occupations. The apocalyptic scenario, in which a complete collapse of social organization ushers in a tumultuous upheaval, is ultimately a mystical rather than a political one. It is much more helpful to think about revolution as a dramatic convergence of real practices of rebellion and resistance that, in their intersection, acquire a qualitatively new form. Interestingly, this emerges as a central theme of one uniquely subversive zombie film, Wes Craven's *The People under the Stairs* (1991).

The People involves a clever blending of racial, class, and gender dynamics via tropes from the horror genre. At its center lies the conflict between white capitalist landlords (known simply as "Daddy" and "Mommy") who are buying up large chunks of ghetto real estate, on the one hand, and on the other hand the zombified white people imprisoned under the couple's cellar stairs and the poor black families from the surrounding community who are tenants in the couple's buildings. It is intriguing to learn that Daddy and Mommy made their initial fortune operating a funeral home, where the bulk of the action takes place, and by selling cheap caskets—profiting, in other words, from the modern corpse economy. Having undertaken this "primitive accumulation," they graduated to dispossessing the living by buying up ghetto real estate.

The social geography of class and race looms large in the film. The funeral home owned by these white capitalist landlords and property developers lies at a remove from the ghetto properties they are buying up, thus underlining the spatiality of class and racial separations. Similarly, the social organization of their (funeral) home is itself spatially stratified, with poor white youth locked under the stairs in their cellar, while the couple moves around in the upper floors. Revolt, when it comes, is thus truly a rebellion from below. Moreover, that revolt is led by a thirteen-year-old black youth intent on fighting an eviction ordered by Daddy and Mommy. Nicknamed "Fool" by his sister—in obvious reference to similar characters of lower-class origin who, in Shakespeare's plays in particular, regularly outwit people of higher social standing—this youngster allies himself with two young whites, one of whom initiates Fool into knowledge of the secret corridors that run inside the walls of the funeral home. Strategically deploying these hidden passages, Fool lays the basis for a liberation struggle—both uprising and exodus—that also seizes the vast supplies of money secreted in a vault inside the house. (Overwhelmed by the immense wealth he sees in the vault, Fool exclaims, "No wonder there's no money in the ghetto.")

First, however, Fool has to win over the zombified white youth kept in captivity and hooked on television images of the U.S. military's bombing of Iraq—a clear reference to the zombifying effects of patriotism on working class whites. As he does that, a sort of guerrilla struggle ensues inside the former funeral home, one that signals antiracist and antipatriarchal role reversal when Fool refers to Daddy as "boy." But, as if debunking mythical notions of rebellion, the film shows us that the zombie uprising can only succeed by means of an alliance with masses of people from the local black community. Fortunately for the zombie-rebels, a community onslaught against Daddy and Mommy is being led by Fool's sister, Ruby, who declares, upon arriving at the landlords' doorstep: "I represent the association of people who have been unduly exploited, evicted, and generally fucked over." This convergence of an uprising of zombified white "insiders" with a growing rebellion of black "outsiders" is decisive. Daddy and Mommy are overthrown, their wealth seized, and "distributed" across the community after Fool dynamites the vault. As insurgent blacks and freed white zombies dance in the streets with money raining down upon them, the movie closes to the sounds of Redhead Kingpin and the FBI singing "Do the Right Thing."

I have dwelled on this film because its combination of discrete zombie imageries critically exposes social dynamics of class and race while also rendering a rebellion of the outcast and the marginalized in everyday terms. The idioms of monstrosity are demonstrably at work here, particularly in the funeral home setting and the idea that Daddy and Mommy both zombify and dispossess poor people. The (white) people under the stairs are immersed in a world of the living-dead, oblivious to the emancipatory possibilities around them. Yet, they possess a latent capacity for awakening, for rising up to reclaim their humanity. In so remaking themselves via their contact with insurgent blacks, they metamorphose into what is arguably the most subversive of monster figures, the zombie rebel. Just as, to paraphrase Marx, the working class must negate its own alienated condition if it is to emancipate itself, so zombie rebels must *de-zombify* themselves and acquire consciousness and identity in the process of overturning their degraded state.[49] Liberation is thus simultaneously the overthrowing of oppressors *and* the self-transformation (the de-zombifying) of the oppressed themselves.

While catastrophic tropes are used in *The People Under the Stairs* and similar stories, they are extracted from apocalyptic scenarios and returned to everyday life. To be a zombie in this idiom is to be an ordinary member of the downtrodden and oppressed majority—it is to be dominated by bosses, bureaucrats and landlords like Daddy and Mommy. In other words, it is the catastrophes of everyday life—mindless toil, racial discrimination, unemployment, eviction, low-wages, insecurity, lack of control over one's life—that are at issue here, as illustrated in Ruby's claim to represent "the unduly exploited, evicted, and generally fucked-over."

For the vast majority in capitalist society—those enmeshed in the living-dead machinery of wage-labor, as well as the thoroughly marginalized—there is a catastrophic texture to everyday life that manifests itself in a life full of "dead-time," insult and humiliation, the degradation of being treated as a beast of burden. The horror idiom partially speaks the truth of these experiences, but frequently in codes and tropes, deployed by literature and film in particular, that separate catastrophe from the world of the everyday. The problem for critical theory and practice is to redeem the truths embedded in monstrous tales while translating them into languages and practices of social and political action. Or, as the radical German critic Walter Benjamin put it, "It is a question of the dissolution of 'mythology' into the space of history."[50]

It follows that a radical theory needs to name the everyday catastrophes that comprise so much of life in capitalist society, while showing their (nonmythical) roots in oppressive but generally mystified social relations. This means reading them dialectically to disclose what they tell us about the genuinely monstrous, deadening, and zombifying processes to which wage-laborers are subjected in modern society. Such a radical reading can thereby reveal the great truth-content of the monster genres that proliferate throughout late capitalism, while also taking them out of the realm of the apocalyptic. We need, in short, to uncover the *social* basis of all that is truly horrifying and catastrophic about our world, as part of a critical theory and practice designed to change it.

Contributors

James Davis is an Irish writer and filmmaker based in California. His documentary films include *Meeting Room*, *Safety Orange*, and *Autonomy and a Song*. He is a contributor to *Confronting Capitalism: Dispatches from a Global Movement*.

Doug Henwood is publisher and editor of *Left Business Observer*. Among other books, he is the author of *After the New Economy* and *Wall Street: How It Works and for Whom*. He is a contributing editor to *The Nation* magazine.

Sasha Lilley is a writer and radio broadcaster. She is the author of *Capital and Its Discontents: Conversations with Radical Thinkers in a Time of Tumult*, series editor of PM Press's political economy imprint, Spectre, and host of *Against the Grain*, the program of radical ideas.

David McNally is professor of political science at York University, Toronto. He is the author of six previous books, including *Against the Market*; *Another World Is Possible: Globalization and Anti-Capitalism*; and *Global Slump: The Economics and Politics of Crisis and Resistance*.

Eddie Yuen teaches urban studies at the San Francisco Art Institute, and is on the editorial board of the journal *Capitalism, Nature, Socialism*. Yuen is also the coeditor, with Daniel Burton-Rose and George Katsiaficas, of *Confronting Capitalism: Dispatches from a Global Movement*. He is working on a book on the political economy of extinction.

Notes

Foreword: Dystopia Is for Losers

1 Michal Kalecki, "Political Aspects of Full Employment," in *Collected Works of Michal Kalecki*, vol. 1, ed. Jerzy Osiatynski (New York: Oxford University Press, 1990), http://mrzine.monthlyreview.org/2010/kalecki220510.html.

2 John Maynard Keynes, "Treatise on Money," in *The Collected Works of John Maynard Keynes* vol. 6, ed. Donald Moggridge (Cambridge: Cambridge University Press, 1978), 259.

3 Frederick Engels, "Outlines of a Critique of Political Economy," http://www.marxists.org/archive/marx/works/1844/df-jahrbucher/outlines.htm.

4 A.R. Ammons, *Sphere: The Form of a Motion* (New York: W.W. Norton, 1974), 30.

5 Data from Angus Maddison's database, http://www.ggdc.net/MADDISON/Historical_Statistics/horizontal-file_02-2010.xls.

6 Ammons, *Sphere*, 29.

Introduction: The Apocalyptic Politics of Collapse and Rebirth

1 The term "catastrophism" comes from science, as a theory of abrupt geologic change over time, in contrast to uniformitarianism, but we are deploying the word politically. As far as we can tell, the word has been used in such a way on the left very sporadically, and variably, as in Perry Anderson's *Considerations on Western Marxism* (London: Verso, 1976) and Antonio Negri's *Marx Beyond Marx: Lessons on The Grundrisse* (New York: Autonomedia, 1996). See also Iain Boal's "Climate, Globe, Capital: The Science and Politics of the Abyss," *SUM Magazine*, December 2009. More generally, there have been several left-wing books written of late about capitalism and apocalypse, including Slavoj Žižek's *Living in the End Times* (New York: Verso, 2010) and Evan Calder Williams, *Combined and Uneven Apocalypse* (Ropley: Zero Books, 2011). The most marvelous, perhaps, is a work of fiction. In Terry Bisson's send-up of the right-wing *Left Behind* novels, all fundamentalists Christians and conservatives are raptured up to heaven, leaving the earth to the left. Communism blooms. *The Left Left Behind* (Oakland: PM Press, 2009).

2 An example of this sentiment can be found in Norman Cohn's examination of millenarian movements, such as the Anabaptists, in *The Pursuit of the Millennium: Revolutionary Millenarians and Mystical Anarchists of the Middle Ages* (Oxford: Oxford University Press, 1970)

3 This book does not have the space to explore liberal catastrophism in depth, but the authors hope that it may encourage others to do so. Political scientist Robert D. Putnam illustrates one variant of liberal catastrophism in his bestselling book *Bowling Alone: The Collapse and Revival of American Community* (2000), which mourns the decline of civic life and its benefits—social capital—in American society. "Creating (or recreating) social capital is no simple task. It would be eased by a palpable national crisis, like war or depression or natural disaster, but for better

and for worse America at the dawn of the new century faces no such galvanizing crisis" (New York: Simon and Schuster, 2000), 402. In his book *Fear: The History of an Idea* ([Oxford: Oxford University Press, 2004], 156), Corey Robin writes of liberals like Michael Ignatieff and Judith Shklar who embraced the "war on terror" following the attacks of September 11, 2001: "Prior to 9/11, some of the most far-reaching theorists of these persuasions had imagined a foreign policy disaster that would deliver the United States from the moral lethargy and creeping despair that allegedly set in after the 1960s, the Cold War, and the triumph of the free market." They viewed the 2001 attacks as revivifying for the national spirit.

4 Naomi Wolf epitomizes such politics in her book *The End of America: A Letter of Warning to a Young Patriot* (White River Junction, VT: Chelsea Green, 2007).

5 Chris Hedges, "Is America 'Yearning for Fascism?'" Truthdig, March 29, 2010, http://www.truthdig.com/report/page2/is_america_yearning_for_fascism_20100329/.

Liberals also have their own wishful thinking about collapse. Kirkpatrick Sale argues, "The collapse [of technocommercial society] will come sooner than we realize—I have predicted within a decade—and it will open up secession (or some equivalent such as city-states or medieval walled cities) as the only possible opportunity for a new society with new human-scale alternatives. I'm not predicting it, mind you, I'm just saying it's the only way to go." Kirkpatrick Sale, "The Decline of the American Empire," *CounterPunch*, January 23, 2012.

6 See particularly Rebecca Solnit's excellent *A Paradise Built in Hell: The Extraordinary Communities That Arise in Disaster* (New York: Viking, 2009).

7 Harry Cleaver, "The Uses of an Earthquake," *Midnight Notes* 9 (May 1988).

8 Three of us are active participants, while two of us are within Retort's wide orbit.

9 T.J. Clark, "For a Left with No Future," *New Left Review* 74 (March–April 2012).

Chapter One: The Politics of Failure Have Failed

1 Stockholm Resilience Center, "The Nine Planetary Boundaries," http://www.stockholmresilience.org/research/researchnews/tippingtowardstheunknown/thenineplanetaryboundaries.4.1fe8f33123572b59ab80007039.html.

2 Even so, the staggering accumulation of dire environmental trends that has mounted in recent years should be enough to make even the most rapacious vulture capitalist take pause. It is important to take a sober look at the contours of projected environmental crises. It's not a pretty picture, but the most remarkable thing about the emerging scientific consensus is the ignorance of the general public, including the political left. This gap between awareness and evidence means that any serious discussion of environmental crisis runs straight into the politics of apocalypse. The question is no longer whether there will be environmental catastrophes, but for whom. To paraphrase William Gibson, "the catastrophe is already here, it's just not evenly distributed."

There are many converging environmental crises today, but all will be amplified if not instigated by the massive atmospheric carbon infusion of the last two centuries of industrial capitalism. The "best case scenario" effects of climate change as it stands today are breathtaking and largely unknown to the publics of the rich world. These include the displacement of millions due to coastal inundation, the salinization of much agricultural land, the "cooking" of Africa, the obliteration of entire ecosystems such as coral reefs, the desertification of the Amazon, the disappearance of the glacial-fed rivers of Asia and South America, the extinction of at least 35 percent of global species, and the advent of the "sea of slime," to name a few.

3 Tom Leonard, "Ready for Doomsday: Buying Asteroid-Proof Bunkers, Killing Their Pets and Planning Mass Suicide, the Families Convinced This Ancient Calendar Predicts the World Will End in 2012," http://www.dailymail.co.uk/news/article-2084476/Doomsday-2012-Meet-families-convinced-Mayan-calendar-predicts-end-world.html.

4 Naomi Wolf, *The End of America: Letter of Warning to a Young Patriot* (White River Junction, VT: Chelsea Green Publishing), 2007.

5 "Increased Knowledge About Global Warming Leads To Apathy, Study Shows," Science Daily, March 27, 2008, http://www.sciencedaily.com/releases/2008/03/080327172038.htm. See also Stefan Skrimshire, "Curb Your Catastrophism," *Red Pepper*, http://www.redpepper.org.uk/curb-your-catastrophism/.

6 David Harvey, *Justice, Nature and the Geography of Difference* (New York: Blackwell Publishers, 1996), 149.

7 John Vidal, "Bill Gates Backs Climate Scientists Lobbying for Large-Scale Geoengineering," *Guardian*, http://www.guardian.co.uk/environment/2012/feb/06/bill-gates-climate-scientists-geoengineering.

8 For more on capitalism and environmental catastrophe see Brian Tokar, "Movements for Climate Action: Toward Utopia or Apocalypse?" *IAS: Perspectives in Anarchist Theory*. See also John Bellamy Foster, "Capitalism and Degrowth: An Impossibility Theorem," *Monthly Review* 62, no. 8 (January 2011): 26–33. http://monthlyreview.org/2011/01/01/capitalism-and-degrowth-an-impossibility-theorem.

9 Peter J. Taylor, "How Do We Know We Have Global Environmental Problems? Undifferentiated Science-Politics and Its Potential Reconstruction," in *Changing Life: Genomes-Ecologies-Bodies-Commodities*, eds. Peter Taylor, Saul Halfon and Paul Edwards, 149–74 (Minneapolis: University of Minnesota Press, 1997).

10 Cindi Katz, "Under the Falling Sky: Apocalyptic Environmentalism and the Production of Nature," in *Marxism in the Postmodern Age*, ed. A. Callari et al. (New York: Guildford, 1995), 277.

11 Video games, it must be remembered, are the dominant cultural form among those under thirty.

12 Tara Andrews and Idit Knaan, "Scared Straight: Don't Believe the Hype (Facts from CJJ)," Reclaiming Futures, January 13, 2011, http://www.reclaimingfutures.org/blog/juvenile-justice-reform-Scared-Straight-Facts-vs-Hype.

13 Frances Fox Piven and Richard A. Cloward, *Why Americans Don't Vote* (New York: Pantheon, 1988).

14 Erik Swyngedouw, "Impossible Sustainability and the Post-political Condition," in *Making Strategies in Spatial Planning, Urban and Landscape Perspectives 9*, edited by Maria Cerreta (New York: Springer, 2010).

15 Significantly, many African Americans and other people of color did not harbor such illusions and therefore did not suffer the "hard landing" of disillusionment to the same extent.

16 John Tirman, "Why Was No One Punished for America's 'My Lai' in Iraq?" Alternet, February 12, 2012, http://www.alternet.org/world/154087/why_was_no_one_punished_for_america's_22_in_iraq?page=entire.

17 Blake Ellis, "Sales of Luxe Doomsday Bunkers up 1,000%," CNN, March 26, 2011. http://money.cnn.com/2011/03/22/real_estate/doomsday_bunkers/index.htm?hpt=C2.

18 Adam Smith, "Global Warming Reopens the Northeast Passage," *Time*, September 17, 2009, http://www.time.com/time/world/article/0,8599,1924410,00.html.

19 "Disaster capitalism," like the terms "capitalist patriarchy" or "racial capitalism" may in this case be redundant, but necessarily so, given the imprecise usage of the word.

20 Mike Davis, *Late Victorian Holocausts: El Niño Famines and the Making of the Third World* (London: Verso, 2002).

21 It is a tribute either to insanity or the eternal optimism of global elites that any silver lining can be seen in this weather forecast. Yet political leaders in Russia, Canada, and other northern nations have expressed enthusiasm for the immanent accessibility of new land masses within their borders, potentially laden with unexploited oil, natural gas, diamonds, tillable soil and other resources. The ocean floors of the polar regions are also exciting frontiers for exploitation, and Canada has significantly beefed up its navy with the intention of patrolling its newly thawed territories, while Russia planted a flag on the ocean floor. In a bizarre expression of utilitarian logic, it is now estimated that the opening of an ice free "northern passage" across the North Pole will result in significant annual carbon reductions.

22 Drew Westen, "What Happened to Obama?" August 7, 2011, http://query.nytimes.com/gst/fullpage.html?res=9C03E7DF1630F934A3575BC0A9679D8B63&pagewanted=all.

23 Hall spoke at UC Santa Cruz in the mid 1990s.

24 Here are some examples of psychic predictions for 2011: http://www.relativelyinteresting.com/psychic-fails-2011-failed-and-forgotten-predictions/.

25 "NASA Space Shuttles Destroy the Ozone Shield," http://www.ringnebula.com/project-censored/1976-1992/1990/1990-story4.htm; "Rockets Destroying Ozone Layer, Say Scientists," *Green Left 10*, http://www.greenleft.org.au/node/1487.

26 Helen Caldicott, "Accidental Armageddon," The Co-Intelligence Institute, http://www.co-intelligence.org/y2k_caldicott.html. Needless to say,

Caldicott is absolutely right to critique both nuclear power and nuclear weapons, but the invocation of immanent doom does not help the cause.

27 "Y2K – Panic Now and Avoid the Rush," Gold Eagle, http://www.gold-eagle.com/editorials_99/mcintosh041599.html.

28 Ken Keyes, Jr., "The 100th Monkey: A Story about Social Change"; Elaine Myers, "The Hundredth Monkey Revisited," The Wow Zone, http://www.wowzone.com/monkey.htm.

29 Eric B. Ross, "The Malthus Factor: Poverty, Politics and Population in Capitalist Development," Corner House Briefing 20, July 31, 2000, http://www.thecornerhouse.org.uk/resource/malthus-factor.

30 Stefan Skrimshire, "Curb Your Catastrophism," *Red Pepper*, http://www.redpepper.org.uk/curb-your-catastrophism/.

31 Haydn Washington and John Cook, *Climate Change Denial: Heads in the Sand* (New York: Routledge, 2011).

32 David F. Noble, "The Corporate Climate Coup," ZNet, May 8, 2007, http://www.zcommunications.org/the-corporate-climate-coup-by-david-f-noble.

33 The forthcoming article "Climate Leviathan" by Joel Wainwright and Geoff Mann has a very interesting analysis of potential state strategies toward climate change.

34 Mike Davis, "Spring Confronts Winter," *New Left Review* 72 (November–December 2011).

35 James O'Connor, *Natural Causes: Essays in Ecological Marxism* (New York: Guilford Press, 1998), 163.

36 Joel Kovel describes carbon trading in the following way in *Enemy of Nature: The End of Capitalism or the End of the World* (London: Zed Books, 2007): "*Kyoto proceeds on a two-tiered front: to create new markets for trading credits to pollute among the industrial powers, and to create* . . . 'Clean Development Mechanisms' . . . in the South that would offset carbon emissions by building projects, like tree farms, whose goal is the sequestration of carbon. This immense superstructure . . . rests on two guiding assumptions: Give the corporate sector and the capitalist state the leading role in containing global warming; and do so by making the control of atmospheric carbon the site of new markets and new nodes of accumulation . . . The defects of this mammoth blunder are myriad. The scheme is inherently incoherent, for it entails innumerable points that simply cannot be measured or compared. This is essentially because it tries to evade the point of a rational policy, which would be to leave the carbon in the ground in the first place—in other words, one that would put limits on capital. In doing so, Kyoto offers opportunities for swindling of all kinds" (47–48).

37 If there is any aspect of the ecological catastrophe that is truly a zero-sum game it is the collapse of biodiversity brought about by what many scientists are calling the sixth great extinction event in the Earth's history. The long-term ecological impoverishment brought about by this crisis is catastrophic in ways that are not yet understood. The logic of extinction is homogenizing: it is the opposite of the mantra of "diversity" which we associate with capitalism since the 1970s. This means less "diversity" to

commodify and less raw material to harvest, patent and subdivide. The unfathomable mysteries of the tropics and the seas will be lost to bioprospecting. The potential cures found from amphibians, snake venom, the herbal and plant knowledge of indigenous humans and other species (such as primates, birds and elephants) will be lost to the pharmaceutical companies. For consumers of the rich world, the discovery of "new" delicacies like goji and açai berries, rooibos tea or the "ornamental multiculturalism" used to market commodities may grind to a halt. Capital valorizes and instrumentalizes the diversity of nature, but this impulse is overwhelmed by its short-term logic of destruction. This contradiction is leaving capital with less and less hope for new "magic bullets" to appear from the outside. Thus, the effort by some sectors of capital to save tropical rainforests is at odds with the short-term imperative to clear these forests for biofuel, palm oil, and soy plantations. In this contest of "interest groups," it is clear that short-term pillaging has greater influence than prudent managerialism at the present time. Neoliberal capitalism has produced neoliberal nature and this has impoverished the capacity for adaptation that may be this epoch's only hope. Understanding this process is not just a matter of social pathology or regulatory failure, though those are certainly factors. The underlying problem is that capital tends to degrade the conditions of its own production, and these kinds of catastrophes are not readily solved by "market solutions." The strip mining of biodiversity should clearly be seen as a catastrophe for the reproduction of capitalism, but it barely registers as a problem in boardrooms or the business press. As it stands, agribusiness concerns, wildlife traffickers and cattle ranchers have more sway than eco-tourism brokers and bioprospecting pharmaceutical companies, and the resulting impoverishment of eco-systems and cultures is a catastrophe that may not be fully recognized until it is too late.

38 Peter Ward, *The Flooded Earth: Our Future in a World Without Ice Caps* (New York: Basic Books, 2012). Note: this prognosis does not factor in the rise of sea levels.

39 Christian Parenti, *Tropic of Chaos: Climate Change and the New Geography of Violence* (New York: Nation Books, 2011).

40 A link to a 2010 report on the most endangered cities: http://www.oecd.org/dataoecd/16/10/39721444.pdf.

41 Pat Speer, "Climate Change: Insurers Confirm Growing Risks, Costs," Insurance Networking News, http://www.insurancenetworking.com/news/insurance-climate-change-risk-ceres-30007-1.html.

42 Parenti, *Tropic of Chaos*, 13–20.

43 Robert D. Kaplan, "Waterworld," *Atlantic*, http://www.theatlantic.com/magazine/archive/2008/01/waterworld/6583/.

44 Kovel, *The Enemy of Nature*, 43; Franz Broswimmer, *Ecocide: A Short History of the Mass Extinction of Species* (London: Pluto, 2002).

45 Eric Swyngedouw, "Apocalypse Forever? Post-political Populism and the Spectre of Climate Change," *Theory, Culture & Society* 27, no. 2–3 (March–May 2010): 213–32. Full text available as pdf at: http://tcs.sagepub.com/content/27/2-3/213.

46 To understand the unprecedented sounds of alarm emitting from the scientific establishment it is important to understand two things. First, the emerging scientific consensus on rapid climate change is truly terrifying. Second, the authority of scientists is being ferociously questioned by an extremely well-funded climate denial lobby (this is not to deny that mainstream climate science is much better funded). Normally staid scientists are finding themselves committed to a somewhat shrill position, given the extremity of the crisis. Many working scientists, sad to say, seem oblivious to the debates that took place within the philosophy of science in the mid-twentieth century and critical science studies from the 1960s on. Flannery, as noted earlier, suggests that the very term "global warming" is too benign: it is not apocalyptic enough. Scientists are driven to communicate what the situation is, but are stymied by the seemingly irrational and incomprehensible resistance to the message.

47 An example of elitist and hectoring climate campaigning was the notorious British commercial called "No Pressure," which provoked a massive outcry for depicting school children being blown up for not fighting climate change: http://www.guardian.co.uk/environment/green-living-blog/2010/oct/04/10-10-activism.

48 James H. Jones, *Bad Blood: The Tuskegee Syphilis Experiment, New and Expanded Edition* (New York: Free Press, 1993).

49 James Lovelock, *The Revenge of Gaia* (New York: Penguin, 2006), 149–50.

50 Chris Carlssen, *Nowtopia: How Pirate Programmers, Outlaw Bicyclists, and Vacant-Lot Gardeners Are Inventing the Future Today* (Oakland: AK Press, 2008). Carlssen makes a strong argument for redefining abundance in ways that are not predicated on exploitation of humans and nature.

51 The website Cooling It! No Hair Shirt Solutions to Global Warming by Gar W. Lipow has some suggestions in this regard http://nohairshirts.com/index.php.

52 Betsy Hartman, *Reproductive Rights and Wrongs: The Global Politics of Population Control* (Boston: South End Press, 1995. See also Janet Biehl, *"Ecology" and the Modernization of Fascism in the German Ultra-right*, http://spunk.org/library/places/germany/sp001630/janet.html.

53 Gray Brechin, "Conserving The Race: Natural Aristocracies, Eugenics, and the U.S. Conservation Movement," *Antipode* 28, no. 3 (July 1996): 229–45.

54 Interview with Carl Anthony, *Race, Poverty & the Environment*, http://urbanhabitat.org/20years.

55 Jeanne Pfaelzer, *Driven Out: The Forgotten War Against Chinese Americans* (Berkeley: University of California Press, 2008).

56 James Howard Kunstler, *The Long Emergency: Surviving the Converging Catastrophes of the Twenty-First Century* (New York: Atlantic Monthly Press, 2005).

57 Fortunately, a compassionate, egalitarian, and radical climate movement already exists. It is called, broadly speaking, the Climate Justice Movement and is evolving on the ground in South America, South Africa, South Asia, and the Pacific, as well as within the United States.

58 Parenti, *Tropic of Chaos*.

Chapter Two: Great Chaos Under Heaven

1 To be clear, when I refer in this essay to politics built around "external" forces causing capitalism's collapse, I am pointing to notions that regard the contradictions of the capitalist system as based on insuperable limits, primarily outside of the realm of the class struggle. I am not suggesting that there are not contradictions within capitalism—on the contrary, there are deep contradictions. I am simply arguing that capitalism will not meet its demise by hitting the wall of these limits.

2 Georg Lukács noted in a different context, "Fatalism and voluntarism are only mutually contradictory to an undialectical and unhistorical mind." He was, of course, not immune to catastrophism. Georg Lukács, *History and Class Consciousness*, (Cambridge: MIT Press, 1972), 4. Such a combination of voluntarism and fatalism can be seen, for example, in the Comintern's Third Period.

3 This chapter also does not examine in any depth radical opponents of left catastrophism. Nor does it focus on thinkers like Walter Benjamin who grappled with the question of catastrophe and the left, eventually concluding a year before his death: "The experience of our generation: that capitalism will not die a natural death." Benjamin, *The Arcades Project*, (Cambridge: Harvard University Press, 2002), 917.

4 Joe Bavier, "Congo War-Driven Crisis Kills 45,000 a Month: Study," *Reuters*, January 22, 2008.

5 Karl Kautsky, *The Class Struggle (Erfurt Program)* (Chicago: Charles H. Kerr, 1910), 117. Kautsky, during his erratic political career, moved from collapsarian to critic of collapse theory to believer that capitalism had transcended crises. See note 12.

6 Joseph A. Schumpeter, *Capitalism, Socialism, and Democracy* (New York: Harper & Row, 1976).

7 *Karl Marx and Frederick Engels, Collected Works* vol. 40 (New York: International Publishers, 1975), 199, 203.

8 The *Grundrisse*, published more than half a century after Marx's death, is often cited to illustrate Marx's collapsism. Marx wrote, "Hence the highest development of productive power together with the greatest expansion of existing wealth will coincide with depreciation of capital, degradation of the labourer, and a most straitened exhaustion of his vital powers. These contradictions lead to explosions, cataclysms, crises, in which by momentaneous suspension of all labor and annihilation of a great portion of capital, the latter is violently reduced to a point where it can go on fully employing its productive powers without committing suicide. Yet, these regularly recurring catastrophes lead to their repetition on a higher scale, and finally to its violent overthrow" (*Grundrisse: Foundations of the Critique of Political Economy* [Harmondsworth: Penguin, 1973], 750.) However, Marx wrote the *Grundrisse* during the 1957 crisis, after which his ideas changed. As John Crump indicates, "The crisis of 1857 and its failure to evoke a revolutionary response from the working class had a big impact on Marx. So when he came to publish *Capital* (Volume I, 1867), although he outlined the cycle of modern industry as 'a series of periods of moderate activity, prosperity, over-production, crisis and stagnation,'

there were no references to revolution automatically arising from this sequence. But if Marx seems to have largely shaken himself free of his former romantic notions, they remained well in evidence in Engels's writings. *Anti-Duhring* (1878) in particular was as outspoken in its commitment to the idea that capitalism would 'collapse' as any of his earlier works had been." "Marx and Engels and the 'Collapse' of Capitalism," *Socialist Standard* (April 1969).

9 A point illustrated by F.R. Hansen in her excellent book *The Breakdown of Capitalism*. "The law of the tendency of the rate of profit to fall, Marx's most substantial theoretical commitment to conceptualization of breakdown, is a law of contradictions. . . . The image that emerges is one of effect and counter-effect, engaged in a mutually dependent process of reciprocity and repetition, in an indeterminate and potentially infinite play of interactions which are similar but different. There is no final indication of full foreclosure at any point, and no final indication of impending breakdown or eventual resolution" (29–30).

10 Karl Marx and Friedrich Engels, *Collected Works, Vol. 4, 1844–45* (London: Lawrence and Wishart, 1975), 93.

11 Hansen, *Breakdown of Capitalism*, 9. Hansen argues that after the death of Marx, his successors embraced either static political economy or Hegelian apocalypticism—both which were rejected or transcended by Marx.

12 As Paul Mattick points out, at the time "the most diverse analyses of crisis were all offered to explain the inevitability of capitalism's decline and the abolition of the system to be effected by political movements evoked by this decline." *Economic Crisis and Crisis Theory* (London: Merlin, 1981), 119. Capitalist breakdown had been the orthodoxy of the SPD under Kautsky, although he ended up modifying his position in debate with Bernstein. At that point, Kautsky argued that Marx never had a theory of breakdown—while omitting his own earlier stance. He had previously argued that "the capitalist social system has run its course; its dissolution is now only a question of time" *The Class Struggle [Erfurt Program]* (Chicago: Charles H. Kerr, 1910), 87, 117. Kautsky later shifted to a theory of chronic stagnation and argued that ever-greater crises would ineluctably lead to revolution—and finally discarded the idea that any of his previous scenarios would come to pass. Heinrich Cunow responded to Bernstein by defending breakdown theory, insisting it originated with Marx, and arguing that dwindling markets will lead to the collapse of the system.

13 Hansen, *The Breakdown of Capitalism*, 31.

14 Rosa Luxemburg, *The Accumulation of Capital: A Contribution to an Economic Explanation of Imperialism* (New York: Routledge, 2003).

15 See Norman Geras, "Rosa Luxemburg: Barbarism and the Collapse of Capitalism," *New Left Review* 82 (November–December 1973): 17–37. Yet as Michael Löwy has indicated, Luxemburg also advocated "accelerating" historical processes, as if there were one unilinear direction that history might take and the role of political organizations was to speed up the time it took to reach its appointment with destiny.

16 Lenin objected to the idea that capitalism would reach insurmountable limits, although he was not opposed to declaring it moribund. He

argued that a new imperial stage of capitalism had taken shape, dominated by finance capital, in which the decay of the system might persist for a long time without imploding. Lenin believed that capitalism would only come to an end due to the concerted efforts of the working class—led of course by the vanguard party—rather than collapse or a series of intensifying crises doing the work of politicizing the masses and setting them into motion. Likewise, Trotsky argued against the idea of a catastrophic collapse: "There is no crisis that can be, by itself, fatal to capitalism. The oscillations of the business cycle only create a situation in which it will be easier, or more difficult, for the proletariat to overthrow capitalism. The transition from a bourgeois society to a socialist society presupposes the activity of living people who are the makers of their own history." "Once Again, Whither France?" *Leon Trotsky on France* (New York: Monad Press, 1979), 79. Yet on the cusp of the outbreak of World War II, he argued that capitalism had reached a terminal crisis. He did not posit that capitalism would breakdown automatically, but took it as inevitable that capitalism would not be able to renew itself. He presumed that its "death agony" would provide the conditions for a revolutionary overthrow of the system (*The Transitional Program for Socialist Revolution: The Death Agony of Capitalism and the Tasks of the Fourth International* [New York: Pathfinder Press, 1974]).

17 Bukharin, in *Imperialism and the Accumulation of Capital*, his 1925 assessment of Luxemburg's argument, wrote: "So much for the 'Theory of Capitalist Collapse' as developed by Rosa Luxemburg. What makes this theory so attractive? Its 'economic determinism' ('objective limits to capitalism,' 'strict outlines of economic laws,' etc.). Further, its (alleged) confirmation by empirical facts (sharpening of the situation as a result of the hunt for markets, periods of catastrophes, 'catastrophical character' of the whole imperialist epoch, etc.). Last—but not least—its 'revolutionary' character." Yet he goes on to write that war will lead to breakdown: "It is a fact that imperialism means catastrophe, that, we have entered into the period of the collapse of capitalism, no less." Rosa Luxemburg and Nikolai Bukharin, *The Accumulation of Capital—An Anti-Critique: Imperialism and the Accumulation of Capital* (New York: Monthly Review Press, 1972).

18 Alan Adler, *Theses, Resolutions and Manifestos of the First Four Congresses of the Third International* (New York: Prometheus Books, 1980), 389.

19 R. Palme Dutt, *Capitalism or Socialism in Britain?* (London: The Blackfriars Press, 1931).

20 Such episodes of insurrectionism alternated with moments of "stabilization" of the capitalist system, during which Stalin instructed Communists around the world to make unsavory cross-class alliances. The Chinese Communist Party was mandated by Moscow to subordinate itself to Chiang Kai-shek's Kuomintang, which ultimately killed tens of thousands of the Chinese Communist Party's members in the Shanghai Massacre. Yet such actions probably had as much to do with Stalin's helmsmanship as the economic determinism of the Comintern.

21 C.L.R. James, *World Revolution 1917–1936: The Rise and Fall of the Communist International* (New York, Humanity Books, 1993), 193.

22 Henryk Grossman, *The Law of Accumulation and Breakdown of the Capitalist System* (London: Pluto Press, 1992), 41–42. It should be noted that Grossman accused Luxemburg of quietism, writing that her theory that capitalism will collapse once it has exhausted all noncapitalist markets "anticipate[s] in theory a situation in which capitalism will be automatically destroyed, although we know that [quoting Lenin] there are no absolutely hopeless situations. Luxemburg thus renders the theory of breakdown vulnerable to the charge of a quietist fatalism in which there is no room for the class struggle." Of course, such a charge was leveled against him, which he strenuously denied, seeing his economic theories as consonant with Leninism. See also M.C. Howard and J.E. King, *A History of Marxian Economics: Volume I, 1883–1929* (Princeton: Princeton University Press, 1989), 316–37.

23 Grossman, *The Law of Accumulation and Breakdown of the Capitalist System*, 85.

24 Grossman's analysis was championed by left communists, despite his membership in the Communist Party of Poland. Paul Mattick, *Economic Crisis and Crisis Theory*. London: Merlin, 1981.

25 Anton Pannekoek, "The Theory of the Collapse of Capitalism (1934)," *Capital and Class* 1, no. 3 (Spring 1977): 59–81.

26 "The theory according to which capitalism has today entered its final crisis also provides a decisive, and simple, refutation of reformism and all Party programs which give priority to parliamentary work and trade union action—a demonstration of the necessity of revolutionary tactics which is so convenient that it must be greeted sympathetically by revolutionary groups." Pannekoek, "Theory of the Collapse."

27 Pannekoek, "Theory of the Collapse."

28 Suh Jae-jeong, "'This Crisis is Fundamental Crisis of Capitalism,' Immanuel Wallerstein," *The Hankyoreh*, January 8, 2009.

29 Immanuel Wallerstein, "The Left, I: Theory and Praxis Once Again" in *The Decline of American Power: The U.S. in a Chaotic World* (New York: The New Press, 2003).

30 Suh Jae-jeong, "'This Crisis is Fundamental."

31 Of those early twentieth century theories of collapse, Panitch and Gindin write: "Marxist crisis theories at the time not only seriously misinterpreted the kind of capitalism developing in the United States, they more generally underestimated the long-term potential for domestic consumption and accumulation within leading capitalist states. This was partly due to their failure to appreciate the extent to which working-class industrial and political organizations then emerging would undermine the thesis of the 'immiseration of the proletariat.' But it was also due to their undeveloped theory of the state, which reduced it to an instrument of capital and underestimate its relative autonomy in relation to both imperial and domestic interventions" (Leo Panitch and Sam Gindin, "Capitalist Crises and the Crisis This Time," *The Crisis This Time: Socialist Register 2011* [New York: Monthly Review Press, 2010], 3.

32 Michael Burawoy, "For a Sociological Marxism: The Complementary

Convergence of Antonio Gramsci and Karl Polanyi," *Politics and Society* 31, no. 2 (June 2003): 205.

33 John A. Hobson, *Imperialism: A Study* (New York: James Pott & Company, 1902).

34 Paul A. Baran and Paul M. Sweezy, *Monopoly Capital: An Essay on the American Economic and Social Order* (New York: Monthly Review Press, 1966).

35 David Riazanov, *Karl Marx and Frederick Engels: An Introduction to Their Lives and Work* (New York: Monthly Review Press, 1974), 67. See also Wilhelm Weitling, *The Poor Sinner's Gospel* (London: Sheed and Ward, 1969). Mikhail Bakunin, who was impressed by Weitling, described his ideas as "a very important and extremely dangerous phenomenon" and that his vision for the future was similar to a "herd of animals brought together by coercion" (Quoted in Mark Leier, *Bakunin* [New York: St Martin's Press, 2006], 107).

36 Sergei Nechaev, *Catechism of a Revolutionist* (London: Active Distribution, 1989).

37 However, *The Communist Manifesto* simultaneously emphasized the social power of workers, brought together by capitalist production.

38 Karl Marx and Frederick Engels, *Collected Works* vol. 9 (New York: International Publishers, 1980), 216. *Wage Labour and Capital* was written in 1847 and published in 1849. Crump points to the difference between Marx and Engels during the former's lifetime—Marx emphasizing the importance of socialist consciousness among the working class as a precondition for revolution, while Engels tended to argue that immiseration was crucial. However, as Crump flags, after Marx's death, Engels stressed the importance of consciousness, writing in the introduction to Marx's *Class Struggles in France*, that "where it is a question of a complete transformation of the social organization, the masses themselves must also be in it, must themselves have grasped what is at stake, what they are going in for with body and soul" (Crump, "Marx and Engels").

39 "Like many others I have believed in my youth that as social conditions became worse, those who suffered so much would come to realise the deeper causes of their poverty and suffering. I have since been convinced that such a belief is a dangerous illusion. . . . There is a pitch of material and spiritual degradation from which a man can no longer rise. Those who have been born into misery and never knew a better state are rarely able to resist and revolt." Rudolf Rocker, *The London Years* (Oakland: AK Press, 2005), 25–26.

40 David Roediger, *Toward the Abolition of Whiteness: Essays on Race, Politics, and Working Class History* (New York: Verso, 1994), 29.

41 Rebecca Solnit, *A Paradise Built in Hell: The Extraordinary Communities That Arise in Disaster* (New York: Viking, 2009). See also scott crow, *Black Flags and Windmills: Hope, Anarchy, and the Common Ground Collective* (Oakland: PM Press, 2011).

42 Rebecca Solnit writes, "[Disaster and crisis can make] obvious an injustice or agenda that was opaque before. Sometimes they do so by generating the circumstances in which people discover each other and thereby a

sense of civil society and collective power. But there is no formula; there are no certainties." *Paradise*, 159–69.

43 Steven J. Ross, *Working-Class Hollywood: Silent Film and the Shaping of Class in America* (Princeton: Princeton University Press, 1999), 13. These numbers are for non-farm employees.

44 David Montgomery, *Workers' Control in America: Studies in the History of Work, Technology, and Labor Struggles* (Cambridge: Cambridge University Press, 1980), 43.

45 Montgomery, *Workers' Control in America*, 122.

46 Irving Bernstein, *The Lean Years, 1920–1933* (Baltimore: Penguin Books, 1966), and *Turbulent Years, 1933–1941* (Boston: Houghton Mifflin, 1970). It should be noted that organizing—by Unemployed Councils, the Communist Party in Alabama, and others—started well before 1934. See "The Housing Question: An Interview with Mike Davis" in *Freedom Now! Struggles for the Human Right to Housing in L.A. and Beyond*, eds. Jordan T. Camp and Christina Heatherton (Los Angeles: Freedom Now Books, 2012), 86.

47 "In the past, we know that episodes of racial violence rose at times of economic distress. So, for instance, lynching rates had been steadily decreasing throughout the 1920s, but then rose again in 1930 at the onset of the Great Depression." Amy Wood quoted in Vijay Prashad, "Hate Politics," *Frontline* 29, no. 8 (April 21–May 4, 2012). See also Wood, *Lynching and Spectacle: Witnessing Racial Violence in America, 1890–1940* (Chapel Hill: University of North Carolina Press, 2009).

48 Cal Winslow, "Rebellion from Below," in *Rebel Rank and File: Labor Militancy and Revolt from Below During the Long 1970s*, eds. Aaron Brenner, Robert Brenner, and Cal Winslow (New York: Verso, 2010), 4. Ironically, this was at a time when the working class was being written off by much of the left. Some radicals, however, argued that the left should take seriously those who had the most collective power, including African American workers in key industries like automobile manufacturing, represented at their most radical by groups like the insurgent Dodge Revolutionary Union Movement and League of Revolutionary Black Workers.

49 Max Elbaum, *Revolution in the Air: Sixties Radicals Turn to Lenin, Mao and Che* (New York: Verso, 2006), 235.

50 Winslow, *Rebel Rank and File*, 8.

51 Benjamin Noys explores the thread of "the worse, the better" in postautonomist thought, rooted in the crisis of the 1970s and capitalist restructuring, and linked to an earlier Marxist politics based on the tendency of the rate of profit to fall. Thinkers like Antonio Negri argued that communism was immanent within capitalism and hence that the worse things became for capitalism, the better they would be for communism. "The implication of [Negri's] work, reflecting on the crisis of Fordism and its 'planner-state', was that communism had already arrived and would need to simply be realized. . . . The uncharitable could say that his own reading of the tendency fell victim to the *failure* of verification, with the defeat of the movement of autonomy and Negri's imprisonment." "Apocalypse, Tendency, Crisis," *Mute Magazine* 2, no. 15 (February 3, 2010).

52 Liberals like Todd Gitlin have their own critique of "worsism," the notion of the worse, the better, which they tend to trot out at election time, urging people to vote for the lesser evil. However, it should be stated clearly that a radical critique of catastrophism has little in common with such a perspective. Liberals presuppose that radical or revolutionary rupture with the status quo is undesirable, whereas this work takes as its starting point the necessity of an end to capitalism. The question that we are posing is whether catastrophism, including the politics of the worse, the better, provides the best avenue to such a transformation.
53 A "Resolution on Fascism" was passed at the Fifth Congress of the Comintern in 1924, which concluded that, "In this epoch of the capitalist crisis . . . and the progressive destruction of the capitalist system . . . fascism ends after its victory in political bankruptcy, its internal contradictions leading to its destruction from within." Nicos Poulantzas, *Fascism and Dictatorship* (London: Verso, 1979), 48.
54 James, *World Revolution 1917–1936*, 354.
55 Resolution of political bureau of the KPD, October 10, 1933, cited in Poulantzas, *Fascism and Dictatorship*, 49.
56 James, *World Revolution 1917–1936*, 331–34.
57 Whether the leaders of the Comintern and the German Communist Party actually believed that they would triumph is, of course, debatable. Eric Hobsbawm, who was a young member of the German Communist Party, wrote: "Even as youthful believers in the inevitability of world revolution we knew, or must have known in the last months of 1932, that it was not going to happen just then. We were certainly not aware that by 1932 the international communist movement had been reduced to almost its lowest point since the establishment of the Comintern [in 1919], but we knew defeat was what faced us in the short run. Not we but someone else was making a bid for power. Indeed, neither the rhetoric nor the practical strategy of the KPD envisaged anything like an imminent takeover." *Interesting Times: A Twentieth Century Life* (New York, Pantheon, 2002), 72.
58 This, of course, carries a different meaning than Marx's statement in *Capital* that "force is the midwife of every old society pregnant with a new one. It is itself an economic power." *Capital* vol. 1 (London: Penguin, 1976), 916.
59 James E. Cronin and Carmen Sirianni, eds., *Work, Community, and Power: The Experience of Labor in Europe and America, 1900–1925* (Philadelphia: Temple University Press, 1983), 23.
60 Corey Robin, *Fear: The History of an Idea* (Oxford: Oxford University Press, 2004), 184.
61 In Germany, the Social Democratic Party split over the war; in Russia, the anarchist Peter Kropotkin came out in support of a victory by Russia, the UK, and France.
62 Beverly Silver, *Forces of Labor: Workers' Movements and Globalization since 1870* (Cambridge: Cambridge University Press, 2003), 140.
63 Gabriel Kuhn, *All Power to the Councils! A Documentary History of the German Revolution of 1918–1919* (Oakland: PM Press, 2012).

64 Eric Hobsbawm, *The Age of Extremes: A History of the World, 1914–1991* (New York: Vintage, 1994), 216.

65 New Communist Movement member Max Elbaum argues that many people in the New Left held this idea but, wary of its controversial nature, fewer would have been willing to put it in print. Conversation with the author, April 19, 2012.

66 Jeremy Varon, *Bringing the War Home: The Weather Underground, The Red Army Faction, and Revolutionary Violence in the Sixties and Seventies* (Berkeley: University of California Press, 2004), 27.

67 Susan Stern, *With the Weathermen: The Journey of a Revolutionary Woman* (Garden City: Doubleday, 1976), 128.

68 From the film *The Weather Underground*, directed by Sam Green and Bill Siegel, 2002. While groups like Weathermen believed they were facing fascism, nonwhite groups, such as the American Indian Movement and the Black Panther Party, were exterminated and repressed on scale that white members of the New Left never faced.

69 While inspired by successful national liberation movements around the world, most particularly the Vietnamese following their Tet Offensive in 1968, they had little faith in working-class whites in the United States. See Harold Jacobs, ed., *Weatherman* (Berkeley: Ramparts Press, 1970), 202.

70 Cathy Wilkerson, *Flying Close to the Sun: My Life and Times as a Weatherman* (New York: Seven Stories Press, 2007), 319–20. Wilkerson wrote in her memoir her reservations about the strategy of creating chaos, but that at the time she mistakenly thought the leadership of Weathermen was consciously planning how to avoid the perils that chaos can bring: "Even then, however, I knew that chaos left a terrifying vacuum of leadership, in which the behavior of human beings under pressure could quickly degenerate into the most random violence. Surely we didn't want that. I assumed the leadership at Flint had thought about this and had a plan to introduce us to weapons and fighting in a way that would avoid this disintegration. They must be making provisions to wage war responsibly." "One rationale, familiar to all young people, was that if you cannot be heard, create chaos. Many, I think, hoped that the rising chaos would move the government to withdraw from Vietnam. Mayhem is the weapon of the powerless, as the French learned in Algeria" (320, 144). Members of the Weather Underground reportedly talked about attacking police in neighborhoods populated by people of color in order to bring down repression, believing it would have a radicalizing effect on the people living there. However, they never actually carried out such attacks (Elbaum, *Revolution in the Air*, 71).

71 Varon, *Bringing the War Home*, 35. Weathermen, the RAF, and various other armed groups in Europe were inspired partially by a tract that was drafted under very different political and social conditions in the Global South. The Brazilian Communist revolutionary Carlos Marighella's *Minimanual of the Urban Guerrilla*, written shortly before police ambushed him in 1969, laid out a formula for armed struggle in cities, in which he asserted that militants should add to the climate of chaos as a means of radicalizing the urban population and galvanizing popular support. He advocated

a strategy in which fighters "must become even more aggressive and active, resorting without pause to sabotage, terrorism, expropriations, assaults, kidnappings, executions, etc. . . . The role of the urban guerrilla, in order to win the support of the population, is to continue fighting, keeping in mind the interests of the people and heightening the disastrous situation within which the government must act." In response, the state would have no option but to ratchet up its repression, further radicalizing bystanders.

Whereas the RAF imagined that full-fledged fascism lurked behind the façade of the German state, the urban guerrilla movements inspired by Marighella in Latin America and Western Asia were typically fighting vicious military dictatorships in nations created by imperial domination. Hence, the motives for such movements cannot be simply compared. That said, in the Global South, the Marighellist strategy appealed to movements without a mass base battling repressive regimes. Yet it did not prove successful. In the main, urban populations did not turn out in support of the guerrillas in response to the chaos and attempts at delegitimating the state, while governments became even more wanton and authoritarian in repressing the fighters. Reflecting on guerrilla struggles in Latin America in the 1960s, Vijay Prashad writes, "The military usurped control of the historical dynamic, destroyed a generation of militants, and took advantage of the situation to foment a decade of military rule." *The Darker Nations: A People's History of the Third World* (New York: The New Press, 2008), 162.

The Maoist Sendero Luminoso, or Shining Path, followed a similar logic of fomenting chaos in its insurgency against the Peruvian state. Its model was not that of urban guerrilla warfare, but Mao's strategy of protracted war by surrounding the city from the country. Sendero wanted to return to an idealized rural past, following the compass of "Gonzalo Thought" of their leader, philosophy professor Abimael Guzmán, the self-professed "fourth sword" of Marxism. (Gonzalo Thought drew from strange sources: Guzmán was an admirer of Rip Van Winkle creator Washington Irving's *Mahomet and His Successors*, published in 1850, which he made militants read to study bringing a new order into being). Sendero escalated its war at the moment when Peru was moving from dictatorship to some form of partial democracy, with the intent of forcing a right-wing coup, which they thought would make conditions more auspicious for revolution. They attempted to create chaos by disrupting the economy and institutions of the Peruvian state, attacking barracks, blowing up power lines, and abducting or killing local politicians and wealthy peasants, as well as trade unionists and other members of the left. The Peruvian government declared martial law and indiscriminately raped, tortured, and killed tens of thousands of people, most of whom had nothing to do with the insurgency.

Sendero extolled the decimation of its opponents as well as the bloodletting of its own cadres. In a rare interview, Guzmán justified the infamous massacre of peasants in the village of Lucanamarca, to avenging the killing of a comrade, stating: "If we were to give the masses a lot of restrictions, requirements and prohibitions, it would mean that deep down we didn't want the waters to overflow. And what we needed was for the waters to overflow, to let the flood rage, because we know that when a river floods its banks it causes devastation, but then it returns to its riverbed." He added, "It turned out as the Chairman had said: the reaction is dreaming when it tries to drown the revolution in blood. They should know they are nourishing it, and this is an inexorable law." "Interview with Chairman Gonzalo," *El Diario*, July 24, 1988. The war and counterinsurgency killed upward of seventy thousand people and displaced six hundred thousand. Although Sendero was not able to win militarily, their insurgency significantly destabilized the country, a context in which President Fujimori was able to carry out an internal coup. He suspended the constitution and authorized the military to kidnap and murder thousands of people, and decapitated the organization by hunting down Sendero's leader Guzmán—and was subsequently reelected for his efforts.

The Naxalites, a disparate grouping of Maoists in India, have frequently followed a similar logic of trying to heighten the contradictions. Charu Mazumdar, founder of the Naxals and first General Secretary of the Communist Party of India (Marxist-Leninist) advocated the "annihilation" of class enemies—landlords and local officials in Bengal, where Naxalism began in 1969—as a means to spur peasants to revolt. He believed that a revolution by annihilation would triumph in India within five years. Mazumdar was later expelled from the party, but the Naxals have fought in different groupings for over four decades, facing a ferocious counterinsurgency in the tribal heartland of central and eastern India. The Indian Marxist Jairus Banaji argues that the Naxalites are motivated by the same logic as Sendero Luminoso. "Abimael Guzmán's idea that the countryside would have to be thrown into chaos, churned up, to create a power vacuum, is a mirror image of the CPI (Maoist) strategy. The idea was to generate terror amongst the population and demonstrate the inability of the state to guarantee the safety of its citizens." Spencer A. Leonard and Sunit Singh, "The Maoist Insurgency in India: An Interview with Jairus Banaji," *Platypus Review*, August 2010.

72 Ulrike Meinhof, "On the Liberation of Andreas Baader" in *The Red Army Faction: A Documentary History. Volume 1: Projectiles for the People*, eds. J. Smith and André Moncourt (Oakland: PM Press, 2009), 368.

73 Ibid., 369–70. However, the members of the RAF did recognize that not all forms of chaos were good for radical aims and that some forms might benefit the right instead. The RAF issued a statement following an attack on the Hamburg train station in 1975, which the state accused the RAF of being behind: "this intelligence service-directed terrorist provocation against the people is meant to increase fear and strengthen the people's identification with the state." "The Bombing of the Hamburg Train Station" in *The Red Army Faction*, 378.

74 In fact, for Bonanno, economic crises no longer occur. Capitalism, he believed, was permanently in a state of chaos: "It is now understood that crises do not exist, not because the world is in perfect order but because, on the contrary, it is in complete disorder. It is continually at the mercy

of turbulence that can either increase or decrease, but cannot be considered a 'crisis' in that it in no way corresponds to 'anomalous' situations but simply to the reality of the economic and social setup. For the capitalists Long Range Planning became obsolete at the beginning of the Seventies. One could say that the parallel concept of 'crisis' still exists for some revolutionaries. The timelapse, as we can see, is considerable." *Let's Destroy Work, Let's Destroy the Economy* (London: Elephant Editions, 2007), 34.

75 Alfredo M. Bonanno, *The Insurrectional Project* (London: Elephant Editions, 2000), 16.

76 Transgressio Legis, "The Passage to Revolution," in *We Are an Image from the Future: The Greek Revolt of December 2008*, eds. A.G. Schwarz, Tasos Sagris, and Void Network (Oakland: AK Press, 2010), 328. Many insurrectionists see their moment of action as a cathartic end in itself. The Greek insurrectionist anarchist group Conspiracy of the Cells of Fire, in a statement accompanying the bombing of the back entrance of Ministry of Macedonia-Thrace in Thessaloniki, declared "Whatever we do, we do because we feel it, and it fills us with meaning." "The Unanimity of the Fearful," in *We Are an Image from the Future*, eds. A.G. Schwarz, Tasos Sagris, and Void Network (Oakland: AK Press, 2010), 330.

77 The Invisible Committee, *The Coming Insurrection* (Los Angeles: Semiotext(e), 2009), 96.

78 Immiseration also plays a similar role for the insurrectionists: "It's sufficient to see how social life suddenly returns to a building suddenly deprived of electricity to imagine what life could become in a city deprived of everything," The Invisible Committee, *Coming Insurrection*, 119.

79 Transgressio Legis, "One Day We Jacked a Fire Engine, Got on the CB Radio, and Said, 'Tonight, You Motherfuckers, We Will Burn You All,'" in *We Are an Image from the Future*, eds. A.G. Schwarz, Tasos Sagris, and Void Network (Oakland: AK Press, 2010), 164.

80 Transgressio Legis, "The Passage to Revolution," in ibid., 327.

81 To be accompanied by the more mundane tactics of squatting buildings and leafleting, in which insurrectionists also engage.

82 Cal Winslow, conversation with the author, April 6, 2012.

83 Anonymous, *At Daggers Drawn with the Existent, Its Defenders, and Its False Critics* (Santa Cruz: Quiver Distro, 2006).

84 Alfredo M. Bonanno, "Why Insurrection?" *Insurrection* 1 (December, 1982): 1.

85 Frantz Fanon, *The Wretched of the Earth* (New York: Grove Press, 2004), 2.

86 Anna Cento Bull, *Italian Neofascism: The Strategy of Tension and the Politics of Non-reconciliation* (Oxford: Berghahn Books, 2007), 7. Former Italian president Francesco Cossiga, who was minister of the interior in the late 1970s, told *Quotidiano Nazionale* in 2008 that his contemporary equivalent "should do what I did when I was secretary of the interior. He should withdraw the police from the streets and the universities, infiltrate the movement with secret (provocateurs) agents, ready to do anything, and, for about ten days, let the demonstrators devastate shops, set fire to cars and lay waste the cities. After which, strengthened by popular consent,

the sound of ambulance sirens should be louder than the police cars. The security forces should massacre the demonstrators without pity, and send them all to hospital." Robert Mancini, "Retribution and Revenge: A Recent Interview by Italy's Former President Sheds Light on One of the Most Secretive Periods of the Country's History" *The Guardian*, November 24, 2008.

87 Ricardo Flores Magón, "Manifesto to the Anarchists of the Entire World and to Workers in General," in *Dreams of Freedom: A Ricardo Flores Magón Reader* (Oakland: AK Press, 2005), 146.

88 "In social terms, collapse refers to a massive reduction or simplification of society. Stratification, specialization, bureaucracy, methods of statist control, the arts, economic coordination and organization, population, and networks of destruction will all be significantly simplified." Kevin Tucker, "Agents of Change: Primal War and The Collapse of Global Civilization," *Species Traitor* 4 (2005): 53.

89 Ran Prieur, "Thinking Through the Fall," *Green Anarchy* 9 (Summer 2002): 1, 12.

90 Kevin Tucker, in "Agents of Chaos" *Green Anarchy* 21 (Fall–Winter 2005–2006): 43.

91 Kevin Tucker, "Agents of Change: Primal War and The Collapse of Global Civilization," *Species Traitor* 4 (2005): 63. "It is anti-political in practice. . . . It is about dismantling power rather than seizing it. That may look like insurrections or it may look like people walking away from civilization. Or it may look like ELF type arsons or armed attacks on key points of the electric power grid that is the lifeblood of civilization." Ibid., 64.

92 Kevin Tucker, "Theses on the Fall of Civilization or, How I Learned to Stop Worrying and Embrace the Coming Collapse," *Species Traitor* 2 (Winter 2002): 30.

93 Andrej Grubačić, conversation with the author, June 3, 2010.

94 Kevin Tucker suggests that targeting the lynchpins of "industrial civilization" would create conditions of chaos favorable to collapse. "A tactic one could propose is focusing one's violent abilities into destroying industrial plants, fiber-optic cables, substations, transformers, or dams which fuel electricity. A tactic like this would halt production and potentially cause favorable social mayhem. Because at least the western world (America, Canada etc.) is completely dependent on electricity." "Agents of Change: Primal War and The Collapse of Global Civilization," *Species Traitor* 4 (2005): 43.

95 Aric McBay, Derrick Jensen, and Lierre Keith, *Deep Green Resistance: Strategy to Save the Planet* (New York: Seven Stories Press, 2011), 439. It is worth noting that despite his criticisms of mainstream environmental groups, Jensen receives money from the Wallace Global Fund, which financially sponsors various liberal environmental groups and has as one of its key priorities population control ("sustainable human population," in its own words).

96 Robert J. Alexander, *International Trotskyism* (Durham and London, 1991), 334.

97 Ibid., 334.

98 Ibid., 663–64.

99 Juan Posadas, *Les Soucoupes Volantes, le processus de la matiere et de l'energie, la science et le socialisme* (Paris: Éditions Réed, 1968).

100 Butterfield, Fox, "Mao Tse-Tung: Father of Chinese Revolution," *New York Times* (September 10, 1976).

101 Of the 1960s generation and the commune experience, Janferie Stone writes, "We had bomb shelter visions of a world that, if poisoned, might begin anew. Humanity, cleansed by 'limited nuclear engagement,' tutored by destruction, might make better choices in such a future." Iain Boal, Janferie Stone, Michael Watts, and Cal Winslow, eds., *West of Eden: Communes and Utopia in Northern California* (Oakland: PM Press, 2012), 174.

102 While some liberals and radicals use the term "ultraleft" to designate such maximalism, I have opted to avoid it. The term is mainly used in the negative sense and muddies as much as it clarifies, not least because there is a current within communism that identifies as "ultraleft," which carries a different meaning in its case.

103 Dan Siegel, correspondence with author, January 2, 2012.

104 Hansen, *The Breakdown of Capitalism*, 138–39.

105 Ibid.

106 E.P. Thompson, *The Making of the English Working Class* (London: Penguin, 1991), 420–28.

107 Quentin Bell, *Virginia Woolf: A Biography, Volumes 1–2* (New York: Harcourt, 1972), 8. Woolf was remarking on her former Anglican father, Leslie Stephens.

108 Raymond Williams argued against the narrow alternatives of triumphalism and pessimism on the left. (*Culture and Materialism: Selected Essays* [London: Verso, 2006], 116.) We are suggesting that such hubristic triumphalism often masks a deep pessimism. Perry Anderson hints at this when writing of postwar Trotskyism, "Triumphalism in the cause of the working class, and catastrophism in the analysis of capitalism, asserted more by will than by intellect, were to be the typical vices of this tradition in its routine forms." Anderson, *Considerations on Western Marxism* (New York: Verso, 1989), 101.

109 Christopher Craig Brittain, *Adorno and Theology* (New York: T&T Clark, 2010), 132.

Chapter Three: At War with the Future

1 *Lou Dobbs Tonight*, CNN, April 14, 2005, http://edition.cnn.com/TRANSCRIPTS/0504/14/ldt.01.html.

2 Mail Foreign Service, "Mexico's Fury as U.S. Border Guard Shoots Dead Boy, 14 . . . for Throwing Stones at Him," *Mail Online*, http://www.dailymail.co.uk/news/article-1285302/Mexico-demands-investigation-U-S-border-agent-shoots-teenager-dead--throwing-stones-him.html.

3 Lourdes Medrano, "Border Deaths for Illegal Immigrants Hit Record High in Arizona Sector," *The Christian Science Monitor*, December 16, 2010, http://www.csmonitor.com/USA/2010/1216/Border-deaths-for-illegal-immigrants-hit-record-high-in-Arizona-sector. According to the ACLU, estimates of the death toll range from 3,861 to 5,607 between 1994 and

2009. See Maria Jimenez, *Humanitarian Crisis: Migrant Deaths at the U.S.–Mexico Border* (San Diego: ACLU of San Diego & Imperial Counties & Mexico's National Commission of Human Rights, 2009).

4 Dean Starkman broke down the Dobbs story in the *Columbia Journalism Review* and confirmed, as others had done, that his numbers were fabrications But he concluded that "there was no real harm done in the end, even to immigrant lepers. How many of them have cable?" "Of Lepers and Lou Dobbs," *Columbia Journalism Review*, May 30, 2007, http://www.cjr.org/the_audit/of_lepers_and_lou_dobbs.php?page=all.

5 It should be noted here that the modern right is a diverse inventory of tendencies, segmented throughout in style and emphasis but sharing important compulsions and inspirations.

6 However, there is an important contradiction between the state's appreciation of the extraparliamentary right as an outside defense of inequality and hierarchy, and the fact that some catastrophist right-wing groups see themselves being at war with the state at the same time.

7 According to its own website, Immigration and Customs Enforcement (ICE) funds a network of more than a hundred detention facilities throughout the United States as of June 2012: ten facilities in the Pacific region of the country, twenty-two in the Mountain zone, thirty-six in the Central region, and thirty-six in the Eastern zone. See http://www.ice.gov/detention-facilities/.

8 Interview with Condoleezza Rice, *CNN Late Edition with Wolf Blitzer*, September 8, 2002. Transcript: http://transcripts.cnn.com/TRANSCRIPTS/0209/08/le.00.html.

9 Liberal and left projects profit less from catastrophism in part because they are inspired less by fear than their right-wing counterparts.

10 John Nelson Darby, *On the Eternity of Punishment and the Immortality of the Soul* (London: R.L. Allan, 1870), 16.

11 Joseph M Canfield, *The Incredible Scofield And His Book* (Portland: Ross House Books, 1988).

12 Bernard McGinn, John J. Collins Stephen J. Stein, *The Continuum History of Apocalypticism* (New York: The Continuum International Publishing Group, 2008), 52.

13 George M. Marsden, *Fundamentalism and American Culture* (Oxford: Oxford University Press, 2006), 90.

14 Paul S. Boyer, *When Time Shall Be No More: Prophesy Belief in Modern American Culture* (Cambridge: Harvard University Press, 1994), 108.

15 Laura Lunger Knoppers, Joan B. Landes, *Monstrous Bodies/Political Monstrosities in Early Modern Europe* (Ithaca: Cornell University Press, 2004), 120.

16 Cited in James Alan Patterson, *Changing Images of the Beast: Apocalyptic Conspiracy Theories in American History* (Wheaton, IL: Evangelical Theological Society, 1988), 455.

17 Ibid., 450.

18 Walter Johnson, ed., *The Papers of Adlai E. Stevenson, Vol. 4* (Boston: Little, Brown, and Co., 1972), 128. Cited in T. Jeremy Gunn, *Spiritual Weapons:*

The Cold War and the Forging of an American National Religion (Westport, CT: Praeger Publishers, 2009).

19 NSC 68, *A Report to the National Security Council* (Washington: 1950).

20 Billy Graham, "God in America," PBS, September 11, 2010, http://www.pbs.org/godinamerica/transcripts/hour-five.html.

21 Ben Bagdikian, *The Media Monopoly* (Boston: Beacon Press, 2000).

22 George Burnham, *Billy Graham: A Mission Accomplished* (Grand Rapids, MI: Fleming H. Ravell Co. 1960), 11

23 Breivik in Andrew Berwick (a.k.a. Anders Breivik), ed., *2083: A European Declaration of Independence* (Oslo: 2011), 17–38.

24 William M. Shea, *The Lion and the Lamb: Evangelicals and Catholics in America* (Oxford: Oxford University Press, 2004), 95.

25 Michael Minnicino, "The New Dark Age, The Frankfurt School and Political Correctness," *Fidelio* (Winter 1992).

26 Ibid.

27 The Free Congress Foundation, *Death of the West: Frankfurt School, Cultural Marxism, Political Correctness* (documentary film).

28 Russ Bellant, *Old Nazis, the New Right, and the Republican Party* (Cambridge: Political Research Associates, 1988), 5.

29 Berwick, *2083*, 17–38.

30 Friedrich Hayek, *The Fatal Conceit* (Chicago: University of Chicago Press, 1988), 130–34.

31 Ayn Rand, *Atlas Shrugged* (New York: Signet, 1957).

32 Leo Strauss, "The Crisis of Our Time," in *The Predicament of Modern Politics*, ed. Howard Spaeth, 41–54 (Detroit: University of Detroit Press, 1964), 47–48.

33 Leo Strauss, "What Can We Learn from Political Theory? New School Lecture, 1942," *The Review of Politics* 69, (South Bend, IN: University of Notre Dame, 2007), 527.

34 Leo Strauss, "Notes on Carl Schmitt: The Concept of the Political," http://archive.org/details/LeoStraussNotesOnCarlSchmittsconceptOfThePolitical.

35 Henry Louis Mencken, *The Philosophy of Friedrich Nietzsche* (Boston: Luce And Company, 1908), 105.

36 Victor Davis Hanson, "Mexifornia: Five Years Later" *City Journal* (Spring 2007), http://www.city-journal.org/html/17_1_mexifornia.html.

37 Victor Davis Hanson, "Do We Want Mexifornia?" *City Journal* (Spring 2002), http://www.city-journal.org/html/12_2_do_we_want.html.

38 Gene Maddaus, "Tim Donnelly's Revolution," *LA Weekly*, September 28, 2010, http://www.laweekly.com/2010-10-28/news/tim-donnelly-s-revolution/3/.

39 Pat Buchanan, *Suicide of a Superpower* (New York: St Martins Press, 2011), 267.

40 Simon Kuper, "The Crescent And The Cross," *Financial Times*, November 10, 2007.

41 Bat Ye'or, *Eurabia: The Euro-Arab Axis* (Cranbury, NJ: Associated University Presses 2005).

42 George Weigel quoted in Matt Carr, "You Are Now Entering Eurabia," *Race & Class* 48, no. 1 (2006): 1–22.

43 Following Breivik's arrest, Peder Jensen revealed to police that Fjordman was his pen name. Jerome Taylor, "Unmasked: The Far-Right Blogger Idolised by Breivik," *Independent*, August 6, 2011, http://www.independent.co.uk/news/world/europe/unmasked-the-farright-blogger-idolised-by-breivik-2332696.html.

44 Dan Ritto, "My Experiences with a Danish Sharia Zone," http://gatesofvienna.blogspot.com/2011/09/my-experiences-with-danish-sharia-zone.html.

45 Fjordman, *The Eurabia Code*, http://chromatism.net/fjordman/eurabiacode.htm.

46 The Antideutsche presented themselves as left-wing and anti-imperialist, yet maintain a right-wing and xenophobic platform.

47 This should not obscure the fact that a significant minority of rightists in the United States and Europe are highly critical of Israel and of U.S./NATO wars in Muslim countries.

48 Associated Press in Washington, "US Military Course Taught Officers 'Islam is the enemy,'" *Guardian*, May 11, 2012, http://www.guardian.co.uk/world/2012/may/11/us-military-course-islam-enemy.

49 In 2002, when Jean-Marie Le Pen, known for four decades as an anti-Semite and fascist, polled 16 percent in the first round of the French Presidential election, one million demonstrated against him throughout France. Five years later, Nicolas Sarkozy overcame Le Pen only after assuming his positions on immigration and crime in the 2007 presidential runoff. In 2011, Le Pen handed over leadership of The National Front (FN) to his daughter Marine Le Pen. She immediately became a favorite television guest and her books enjoyed prominent display at the Carrefour checkout. On becoming leader, she announced her intention to fight the Islamic "caliphate" that France was becoming, along with capitalist globalization. Le Pen's populist renovation of the party's manifesto, an autarkic critique of the EU, combined with promises on pensions and employment, is a peculiar mix of purloined left shibboleths and a kind of mystified France first self-sufficiency. See Robert Marquand, "France's Marine Le Pen Aims to Shape a 21st Century Far Right," *Christian Science Monitor*, January 17, 2011.

50 Henry Samuel, "National Front's Marine Le Pen to Prove Formidable Rival to Nicolas Sarkozy," *Telegraph*, December 12, 2010.

51 Ibid.

52 Most of these parties enjoy considerable electoral support and those who are not in coalition governments are often the major voice in opposition. As their influence has grown, ethnic, religious and racist scapegoating has made anti-immigrant attitudes culturally acceptable and increasingly normal. Sarkozy's banning of the burqa to defend his right flank against the republicanized FN and Cameron's eulogy for multiculturalism in response to threats from the British National Party and the UK Independence Party are alarming if only because the hard right is again setting the European agenda. In the 1980s and 1990s, such groups were

quickly branded as fascists and marginalized and their isolation precluded any electoral impact.

53 Ben Berkowitz and Gilbert Kreijer, "Dutch Close in on Government with Anti-Islam Party Backing," June 30, 2010, http://www.reuters.com/article/2010/07/30/us-dutch-politics-idUSTRE66T59E20100730.

54 Geert Wilders, "A Warning to America," Speech given at Cornerstone Church, Nashville, Tennessee, May 12, 2011, http://www.jihadwatch.org/2011/05/geert-wilders-a-warning-to-america.html.

55 Geert Wilders, *Fitna* (Pimpernel Productions, 2008).

56 BBC News, "Denmark's Immigration Issue," February 19, 2005, http://news.bbc.co.uk/2/hi/europe/4276963.stm.

57 The Schengen Area refers to an agreement between twenty-six European countries that operate external border controls but none internally. Only travelers entering and leaving the area pass through border controls.

58 Harriet Alexander, "Denmark's Defiance over Frontier Controls Has Left European Union Bordering on Crisis," *Telegraph*, February 12, 2012.

59 In February of 2012, Muslims in New York called for the resignation of police commissioner Raymond Kelly following the scandal and extended cover-up over the screening of the Eurabia-inspired documentary *The Third Jihad* for 1,500 New York Police Department officers. According to the *New York Times*, the film had played on continuous loop for officers attending antiterrorist training for a year. A contractor for The Department of Homeland Security provided it to the NYPD. The film features interviews with Kelly, Melanie Phillips, and Ayaan Hirsi Ali, alongside a slew of security state insiders and Eurabia-influenced right-wing notables. It argues that Islamic jihad is being built inside the American Gulag where incarcerated and alienated young blacks are drawn to visiting Muslim Imams who are "turning them into radical Islamists." For an American audience, the film then turns to a sequence ruminating on the supposed population time bomb in Europe:

MARK STEYN: "Essentially we live in a world where the most advanced societies are going out of business. In Spain, you have an upside-down family tree. You have four grandparents who have two children who have one grandchild. You do that for another twenty, thirty, forty years, ah, there are not going to be any more Spaniards."

BERNARD LEWIS: "Most of the countries in Western Europe are in the process of becoming Muslim majority countries. The proportion of Muslims in the population is increasing steadily."

MUAMAR GHADAFFI: "Allah is enabling an Islamic nation, which is the Turkish Nation, to join the EU. Fifty million Muslims in Europe will transform Europe into an Islamic continent in Decades."

MARK STEYN: "They do see the fertility rate as a key element of conquest."

BERNARD LEWIS: "Europe is already a lost cause and as time goes on America will be endangered."

See also: Michael Powell, "In Shift, Police Say Leader Helped with Anti-Islam Film and Now Regrets It," *New York Times*, January 24, 2012; and Melanie Phillips, interviewed in *The Third Jihad* (PublicScope Films, 2008).

60 Thomas Hobbes, *Leviathan* (London: Routledge, 1887), 87.

61 Max Horkheimer, Theodor W. Adorno, and Gunzelin Schmid Noerr, eds., *Dialectic of Enlightenment*, (Stanford, CA: Leland Stanford Junior University, 2002), 23.

62 "Established in the spring of 1997, the Project for the New American Century is a non-profit, educational organization whose goal is to promote American global leadership. The Project is an initiative of the New Citizenship Project (501c3); the New Citizenship Project's chairman is William Kristol and its president is Gary Schmitt" (http://www.newamericancentury.org).

63 G.M. Gilbert, *Nuremberg Diary*, (New York: Farrar, Strauss and Company 1947), 279.

64 The state has and continues to use its power to repress right-wing formations and groups that it sees getting out of hand so while there is a symbiosis between the state and the hard right it cannot be argued that the state always overlooks the right as a political threat. "In the 1980s, when a section of the neo-Nazi movement literally declared war on the U.S. government, there were shootouts, sweeps and roundups, conspiracy trials, etc. The U.S. far right does have its own martyrs, such as Gordon Kahl (Posse Comitatus member, killed 1983), Robert Mathews (leader of The Order, killed 1984), and Vicki Weaver and her 14-year-old son Sam (part of a Christian Identity/survivalist family, killed by federal agents during the 1991 Ruby Ridge siege that was as much as trigger for the rise of the militia movement as Waco was). The repression against the militia movement itself was less severe because most militia groups were just defensive formations, but a lot of them were broken up or disrupted with illegal weapons charges and so forth. And I would argue that there is serious potential for a resurgence of significant right-wing revolutionary forces aiming to overthrow (or, more likely, secede from) the U.S. government. I think there are forces in the state that are aware of this potential, and they are not likely to play nice." Matthew Lyons, e-mail to author, June 24, 2012.

65 Richelieu, quoted in Cory Robin, *The Reactionary Mind* (Oxford: Oxford University Press, 2011), 187.

66 NSC 68, *Report*, 3.

67 Ibid., 4.

68 Ibid.

69 William R. Keylor, *The Twentieth-Century World: An International History* (New York: Oxford University Press, 1992), 261–95.

70 Carl Schmitt, *The Concept of the Political* (Chicago: University of Chicago Press, 1996), 27.

71 Giorgio Agamben, "The State of Emergency," lecture at the Centre Roland-Barthes (Université Paris VII, Denis-Diderot, 2002).

72 Associated Press, "Transcript of Treasury Secretary Paulson's News Conference," September 19, 2008.

73 C-Span, October 2, 2008, viewable at YouTube, accessed August 28, 2012, http://www.youtube.com/watch?v=HaG9d_4zij8.

74 These right-wing narratives raise questions about whether the conception of catastrophe as a disease or the cure represents two distinct ideological

forms of the right, or if they are rhetorical strategies deployed in different contexts.

75 McVeigh was convicted of causing the Oklahoma City bombing in 1995, where a truck bomb exploded outside the Alfred Murrah Federal Building killing 168 people including 19 children. The bombing injured more than 600 others. McVeigh and two others were convicted, and he was executed in 2001. McVeigh was involved with the right-wing militia movement and had ties to the white supremacist right. The timing of the bombing may have been related to the execution of one of the leaders of The Covenant, The Sword, and the Arm of the Lord on the same day as the bombing. Richard Wayne Snell had been convicted of murdering a police officer and the owner of a pawnshop. The Covenant was a Christian Identity and white supremacist doomsday cult obsessed with the Federal Government, whom they labeled the Zionist Occupation Government (ZOG). See "White Supremacist Executed for Murdering 2 in Arkansas," *New York Times*, April 21, 1995, http://www.nytimes.com/1995/04/21/us/white-supremacist-executed-for-murdering-2-in-arkansas.html.

76 Samuel P. Huntington, "The Clash of Civilizations?" *Foreign Affairs* 72 (Summer 1993): 22–49.

77 Mark Steyn, *After America* (Washington, DC: Regnery Publishing, 2011).

Chapter Four: Land of the Living Dead

1 Walter Benjamin, *The Arcades Project*, trans. Howard Eiland and Kevin McLaughlin (Cambridge, MA: Belknap Press of Harvard University Press, 1999), 473.

2 To take just one recent example, the author of an interesting book on money and modern society proclaims that we are heading toward an economic and environmental "collapse" in which global population will "plummet toward preindustrial levels over the next thirty years," producing "catastrophes and abominations of hitherto unimagined proportions." See Philip Goodchild, *The Theology of Money* (Durham: Duke University Press, 2009), 66, 48.

3 Lev Grossman, "Zombies Are the New Vampires," *Time*, April 9, 2009. At that very moment, the spoofy novel, *Pride and Prejudice and Zombies* was rocketing up bestseller lists. See Jane Austen and Seth Grahame-Smith, *Pride and Prejudice and Zombies* (Philadelphia: Quirk Books, 2009).

4 W. Scott Poole, *Monsters in America: Our Historical Obsession with the Hideous and the Haunting* (Waco: Baylor University Press, 2011), 14. Poole's study is quite useful, but it fails to take up the link between capitalism, corporeal vulnerability and body panics that I address here.

5 David Graeber, *Debt: The First 5,000 Years* (Brooklyn: Melville House Publishing, 2011), 149.

6 See Susan Tyler Hitchcock, *Frankenstein: A Cultural History* (New York: W.W. Norton, 2007).

7 In what follows I draw upon my own analysis in *Monsters of the Market: Zombies, Vampires and Global Capitalism* (Leiden: Brill, 2011; and in paperback, Chicago: Haymarket, 2012), chap. 1.

8 Mary Shelley, *Frankenstein*, revised edition of 1831 (Toronto: Broadview, 1999), vol. 1, chap. 3, 79.

9 Ibid., 82.

10 Peter Linebaugh, "The Tyburn Riot Against the Surgeons," in *Albion's Fatal Tree: Crime and Society in Eighteenth-Century England*, eds. Douglas Hay et al. (New York: Pantheon Books, 1975).

11 Ruth Richardson, *Death, Dissection and the Destitute* (London: Routledge, 1987).

12 Karl Marx, *Wage Labour and Capital* (Moscow: Progress Publishers, 1952), 20.

13 This point is made nicely from a somewhat different angle by Annalee Newitz, *Pretend We're Dead: Capitalist Monsters in American Pop Culture* (Durham: Duke University Press, 2006), especially pages 6 and 34.

14 James Doherty, *Poor Man's Advocate*, September 1, 1832.

15 This is one point at which a radical psychoanalytic perspective would intervene to explore the social roots of Freud's ostensible "death instinct," recognizing that capitalism eroticizes the dead—commodities—while deadening the living. Walter Benjamin was one of the most important theorists to interrogate this from the standpoint of Marx's account of commodity fetishism, as I discuss in *Bodies of Meaning: Studies on Language, Labor and Liberation* (Albany: State University of New York Press, 2001), chap. 5.

16 Franco Moretti, *Signs Taken for Wonders* (London: Verso, 1983), 85.

17 On film versions of *Frankenstein* see Paul O'Flinn, "On Production and Reproduction: The Case of Frankenstein," *Literature and History* 9, no. 2 (1983): 194–213; and Andrew Milner, *Literature, Culture and Society* (London: UCL Press, 1996), 161–67.

18 As quoted by Hitchcock, *Frankenstein*, 173.

19 Maximilien Laroche, "The Myth of the Zombi" in *Exile and Tradition: Studies in African and Caribbean Literature*, ed. Rowland Smith (New York: Africana Publishing, 1976), 46–48. See also Kevin Alexander Boon, "Ontological Anxiety Made Flesh: The Zombie in Literature, Film and Culture" in *Monsters and the Monstrous: Myths and Metaphors of Enduring Evil*, ed. Niall Scott (Amsterdam, NY: Rodopi, 2007), 36.

20 Melville Herskovits, *Dahomey, a West African Kingdom* (New York: J.J. Augustin, 1938), 243.

21 Laurent Dubois, *Avengers of the New World: The Story of the Haitian Revolution* (Cambridge, MA: Harvard University Press, 2004), 30.

22 On the period of U.S. occupation of Haiti see Laurent Dubois, *Haiti: The Aftershocks of History* (New York: Metropolitan Books, 2012).

23 Joan Dayan, *Haiti, History, and the Gods* (Berkeley: University of California Press, 1995), 37.

24 William Seabrook, *The Magic Island* (New York: Blue Ribbon Books, 1929), 94–95.

25 Ibid., 101.

26 Alfred Métraux, *Le Vaudou Haitien* (Paris: Gallimard, 1957), 250–51.

27 Karl Marx, *The Poverty of Philosophy* (New York: International Publishers, 1963), 54.

28 Orlando Patterson, *Slavery and Social Death* (Cambridge, MA: Harvard

University Press, 1982). See also G.M. James Gonzalez, "Of Property: On 'Captive' 'Bodies,' Hidden 'Flesh' and Colonization in *Existence in Black: An Anthology of Black Existential Philosophy*, ed. Lewis R. Gordon (New York: Routledge, 1997), 129–33.

29 I leave aside here the possibility of the slave being restored to free status, something that was more common in parts of the ancient world than under New World slavery.

30 Laroche, "Myth of the Zombi," 56.

31 As I show in *Monsters of the Market*, chap. 3.

32 While this identification can be found earlier, it is Romero's film that inaugurated a new genre.

33 Robin Wood, "An Introduction to the American Horror Film" in *Movies and Methods* 2, ed. Bill Nichols (Berkeley: University of California Press, 1985), 213.

34 Elliot Stein, "The Dead Zones: George A. Romero at the American Museum of the Moving Image," *Village Voice*, January 8–14, 2003.

35 Gilles Deleuze and Félix Guattari, *Anti-Oedipus: Capitalism and Schizophrenia* (Minneapolis: University of Minnesota Press, 1983), 335.

36 Of course, there are always continuities between older and newer cultural imaginaries, notwithstanding marked transformations, as Luise White points out in her excellent book *Speaking with Vampires: Rumor and History in Colonial Africa* (Berkeley: University of California Press, 2000).

37 As I explain in *Monsters of the Market* (chap. 3), like most critical anthropologists, I use the term "witchcraft" because this is the expression that is widely deployed across sub-Saharan Africa to describe magical and mysterious happenings, especially those related to economic gain.

38 On patterns of rural displacement in Africa, see Pauline E. Peters, "Inequality and Social Conflict Over Land in Africa," *Journal of Agrarian Change* 4, no. 3 (2004): 269–315; and Philip Woodhouse, ed., *African Enclosures? The Social Dynamics of Wetlands in Drylands* (Trenton, N.J.: Africa World Press, 2001). On sub-Saharan Africa's explosive patterns of urbanization see the account by Mike Davis, *Planet of Slums* (London: Verso, 2006). For the disastrous pattern of "development" on the African subcontinent in the neoliberal period see my *Monsters of the Market*, 219–21.

39 Todd Sanders, "Invisible Hands, Visible Goods: Revealed and Concealed Economies in Millennial Tanzania" in *Transparency and Conspiracy*, eds. Harry G. West and Todd Sanders (Durham: Duke University Press, 2003), 164–66.

40 Rosalind Shaw, "The Production of Witchcraft/Witchcraft as Production: Memory, Modernity and the Slave Trade in Sierra Leone," *American Ethnologist* 24, no. 4 (1997): 856–76.

41 Cyprian F. Fisiy and Peter Geschière, "Witchcraft, Development and Paranoia in Cameroon" in *Magical Interpretations, Material Realities: Modernity, Witchcraft and the Occult in Postcolonial Africa*, eds. Henrietta Moore and Todd Sanders (New York: Routledge, 2001), 232–33, 241, 242.

42 Cyprian F. Fisiy and Peter Geschière, "Sorcery, Witchcraft and Accumulation: Regional Variations in South and West Cameroon," *Critique of Anthropology* 11, no. 3 (1991): 255, 260–62; Michael Rowlands and

Jean-Pierre Warnier, "Sorcery Power and the Modern State in Cameroon," *Man* 23, no. 1 (1988): 129.

43 Jean Comaroff and John Comaroff, "Occult Economies and the Violence of Abstraction: Notes from the South African Postcolony," *American Ethnologist* 26, no. 2 (1999): 289. On Tanzania, see Todd Sanders, "Save Our Skins: Structural Adjustment, Morality and the Occult" in Moore and Sanders, *Magical Interpretations*.

44 Peter Geschière, "Globalization and the Power of Indeterminate Meaning: Witchcraft and Spirit Cults in Africa and East Asia" in *Globalization and Identity: Dialectics of Flow and Closure*, eds. Peter Geschière and Birgit Meyer (Oxford: Blackwell Publishers, 1999), 232.

45 World Bank, *World Development Finance* (Washington: World Bank, 2001).

46 Jean Nanga, "The Marginalization of sub-Saharan Africa," *International Viewpoint* 355 (December 2003), at http://www.internationalviewpoint.org/spip.php?article115.

47 See my *Global Slump: The Economics and Politics of Crisis and Resistance* (Oakland: PM Press, 2011), and many of the essays in *The Crisis This Time: Socialist Register 2011*, eds. Leo Panitch, Greg Albo, and Vivek Chibber (London: Merlin Press, 2010), and *The Crisis and the Left: Socialist Register 2012*, eds. Leo Panitch, Greg Albo, and Vivek Chibber (London: Merlin Press, 2011).

48 See my *Another World Is Possible: Globalization and Anti-Capitalism* (Winnipeg: Arbeiter Ring Publishing, 2006), 48.

49 This is also a theme of George A. Romero's films *Day of the Dead* and *Land of the Dead*, in which the zombies appear to be capable of learning and acquiring technological knowledge (how to use guns).

50 Benjamin, *Arcades Project*, 458.

Index

ABOUT PM PRESS

PM Press was founded at the end of 2007 by a small collection of folks with decades of publishing, media, and organizing experience. PM Press co-conspirators have published and distributed hundreds of books, pamphlets, CDs, and DVDs. Members of PM have founded enduring book fairs, spearheaded victorious tenant organizing campaigns, and worked closely with bookstores, academic conferences, and even rock bands to deliver political and challenging ideas to all walks of life. We're old enough to know what we're doing and young enough to know what's at stake.

We seek to create radical and stimulating fiction and non-fiction books, pamphlets, T-shirts, visual and audio materials to entertain, educate and inspire you. We aim to distribute these through every available channel with every available technology — whether that means you are seeing anarchist classics at our bookfair stalls; reading our latest vegan cookbook at the café; downloading geeky fiction e-books; or digging new music and timely videos from our website.

PM Press is always on the lookout for talented and skilled volunteers, artists, activists and writers to work with. If you have a great idea for a project or can contribute in some way, please get in touch.

PM Press
PO Box 23912
Oakland, CA 94623
www.pmpress.org

FRIENDS OF PM PRESS

These are indisputably momentous times—the financial system is melting down globally and the Empire is stumbling. Now more than ever there is a vital need for radical ideas.

In the four years since its founding—and on a mere shoestring—PM Press has risen to the formidable challenge of publishing and distributing knowledge and entertainment for the struggles ahead. With over 175 releases to date, we have published an impressive and stimulating array of literature, art, music, politics, and culture. Using every available medium, we've succeeded in connecting those hungry for ideas and information to those putting them into practice.

Friends of PM allows you to directly help impact, amplify, and revitalize the discourse and actions of radical writers, filmmakers, and artists. It provides us with a stable foundation from which we can build upon our early successes and provides a much-needed subsidy for the materials that can't necessarily pay their own way. You can help make that happen—and receive every new title automatically delivered to your door once a month—by joining as a Friend of PM Press. And, we'll throw in a free T-shirt when you sign up.

Here are your options:

- **$25 a month** Get all books and pamphlets plus 50% discount on all webstore purchases
- **$40 a month** Get all PM Press releases (including CDs and DVDs) plus 50% discount on all webstore purchases
- **$100 a month** Superstar—Everything plus PM merchandise, free downloads, and 50% discount on all webstore purchases

For those who can't afford $25 or more a month, we're introducing **Sustainer Rates** at $15, $10 and $5. Sustainers get a free PM Press T-shirt and a 50% discount on all purchases from our website.

Your Visa or Mastercard will be billed once a month, until you tell us to stop. Or until our efforts succeed in bringing the revolution around. Or the financial meltdown of Capital makes plastic redundant. Whichever comes first.

Also from SPECTRE from PM Press

Capital and Its Discontents: Conversations with Radical Thinkers in a Time of Tumult

Sasha Lilley

ISBN: 978-1-60486-334-5
$20.00 320 pages

Capitalism is stumbling, empire is faltering, and the planet is thawing. Yet many people are still grasping to understand these multiple crises and to find a way forward to a just future. Into the breach come the essential insights of *Capital and Its Discontents*, which cut through the gristle to get to the heart of the matter about the nature of capitalism and imperialism, capitalism's vulnerabilities at this conjuncture—and what can we do to hasten its demise. Through a series of incisive conversations with some of the most eminent thinkers and political economists on the Left—including David Harvey, Ellen Meiksins Wood, Mike Davis, Leo Panitch, Tariq Ali, and Noam Chomsky—*Capital and Its Discontents* illuminates the dynamic contradictions undergirding capitalism and the potential for its dethroning. At a moment when capitalism as a system is more reviled than ever, here is an indispensable toolbox of ideas for action by some of the most brilliant thinkers of our times.

"These conversations illuminate the current world situation in ways that are very useful for those hoping to orient themselves and find a way forward to effective individual and collective action. Highly recommended."
— Kim Stanley Robinson, *New York Times* bestselling author of the *Mars Trilogy* and *The Years of Rice and Salt*

"In this fine set of interviews, an A-list of radical political economists demonstrate why their skills are indispensable to understanding today's multiple economic and ecological crises."
— Raj Patel, author of *Stuffed and Starved* and *The Value of Nothing*

"This is an extremely important book. It is the most detailed, comprehensive, and best study yet published on the most recent capitalist crisis and its discontents. Sasha Lilley sets each interview in its context, writing with style, scholarship, and wit about ideas and philosophies."
— Andrej Grubačić, radical sociologist and social critic, co-author of *Wobblies and Zapatistas*

Also from SPECTRE from PM Press

Global Slump: The Economics and Politics of Crisis and Resistance

David McNally

ISBN: 978-1-60486-332-1
$15.95   176 pages

Global Slump analyzes the world financial meltdown as the first systemic crisis of the neoliberal stage of capitalism. It argues that—far from having ended—the crisis has ushered in a whole period of worldwide economic and political turbulence. In developing an account of the crisis as rooted in fundamental features of capitalism, *Global Slump* challenges the view that its source lies in financial deregulation. It offers an original account of the "financialization" of the world economy and explores the connections between international financial markets and new forms of debt and dispossession, particularly in the Global South. The book shows that, while averting a complete meltdown, the massive intervention by central banks laid the basis for recurring crises for poor and working class people. It traces new patterns of social resistance for building an anti-capitalist opposition to the damage that neoliberal capitalism is inflicting on the lives of millions.

"In this book, McNally confirms—once again—his standing as one of the world's leading Marxist scholars of capitalism. For a scholarly, in depth analysis of our current crisis that never loses sight of its political implications (for them and for us), expressed in a language that leaves no reader behind, there is simply no better place to go."
— Bertell Ollman, professor, Department of Politics, NYU. and author of *Dance of the Dialectic: Steps in Marx's Method*

"David McNally's tremendously timely book is packed with significant theoretical and practical insights, and offers actually-existing examples of what is to be done. Global Slump urgently details how changes in the capitalist space-economy over the past 25 years, especially in the forms that money takes, have expanded wide-scale vulnerabilities for all kinds of people, and how people fight back. In a word, the problem isn't neoliberalism—it's capitalism."
— Ruth Wilson Gilmore, University of Southern California and author, *Golden Gulag*

Also from SPECTRE from PM Press

In and Out of Crisis: The Global Financial Meltdown and Left Alternatives

Greg Albo, Sam Gindin, Leo Panitch

ISBN: 978-1-60486-212-6
$13.95 144 pages

While many around the globe are increasingly wondering if another world is indeed possible, few are mapping out potential avenues—and flagging wrong turns—en route to a post-capitalist future. In this groundbreaking analysis of the meltdown, renowned radical political economists Albo, Gindin, and Panitch lay bare the roots of the crisis, which they locate in the dynamic expansion of capital on a global scale over the last quarter century—and in the inner logic of capitalism itself. With an unparalleled understanding of the inner workings of capitalism, the authors of *In and Out of Crisis* provocatively challenge the call by much of the Left for a return to a largely mythical Golden Age of economic regulation as a check on finance capital unbound. They deftly illuminate how the era of neoliberal free markets has been, in practice, undergirded by state intervention on a massive scale. In conclusion, the authors argue that it's time to start thinking about genuinely transformative alternatives to capitalism—and how to build the collective capacity to get us there. *In and Out of Crisis* stands to be the enduring critique of the crisis and an indispensable springboard for a renewed Left.

"Once again, Panitch, Gindin, and Albo show that they have few rivals and no betters in analyzing the relations between politics and economics, between globalization and American power, between theory and quotidian reality, and between crisis and political possibility. At once sobering and inspiring, this is one of the few pieces of writing that I've seen that's essential to understanding—to paraphrase a term from accounting—the sources and uses of crisis. Splendid and essential."
— Doug Henwood, *Left Business Observer*, author of *After the New Economy* and *Wall Street*

"Mired in political despair? Planning your escape to a more humane continent? Baffled by the economy? Convinced that the Left is out of ideas? Pull yourself together and read this book, in which Albo, Gindin, and Panitch, some of the world's sharpest living political economists, explain the current financial crisis—and how we might begin to make a better world."
— Liza Featherstone, author of *Students Against Sweatshops* and *Selling Women Short: The Landmark Battle for Workers' Rights at Wal-Mart*

Also from SPECTRE CLASSICS from PM Press

William Morris: Romantic to Revolutionary

E.P. Thompson
with a foreword by Peter Linebaugh

ISBN: 978-1-60486-243-0
$32.95 880 pages

William Morris—the great 19th century craftsman, architect, designer, poet and writer—remains a monumental figure whose influence resonates powerfully today. As an intellectual (and author of the seminal utopian *News From Nowhere*), his concern with artistic and human values led him to cross what he called the 'river of fire' and become a committed socialist—committed not to some theoretical formula but to the day by day struggle of working women and men in Britain and to the evolution of his ideas about art, about work and about how life should be lived. Many of his ideas accorded none too well with the reforming tendencies dominant in the Labour movement, nor with those of 'orthodox' Marxism, which has looked elsewhere for inspiration. Both sides have been inclined to venerate Morris rather than to pay attention to what he said. Originally written less than a decade before his groundbreaking *The Making of the English Working Class*, E.P. Thompson brought to this biography his now trademark historical mastery, passion, wit, and essential sympathy. It remains unsurpassed as the definitive work on this remarkable figure, by the major British historian of the 20th century.

"Two impressive figures, William Morris as subject and E. P. Thompson as author, are conjoined in this immense biographical-historical-critical study, and both of them have gained in stature since the first edition of the book was published… The book that was ignored in 1955 has meanwhile become something of an underground classic—almost impossible to locate in second-hand bookstores, pored over in libraries, required reading for anyone interested in Morris and, increasingly, for anyone interested in one of the most important of contemporary British historians… Thompson has the distinguishing characteristic of a great historian: he has transformed the nature of the past, it will never look the same again; and whoever works in the area of his concerns in the future must come to terms with what Thompson has written. So too with his study of William Morris."
— Peter Stansky, *The New York Times Book Review*

"An absorbing biographical study… A glittering quarry of marvelous quotes from Morris and others, many taken from heretofore inaccessible or unpublished sources."
— Walter Arnold, *Saturday Review*

Theory and Practice: Conversations with Noam Chomsky and Howard Zinn (DVD)

Noam Chomsky, Howard Zinn, Sasha Lilley

ISBN: 978-1-60486-305-5
$19.95 105 minutes

Two of the most venerable figures on the American Left—Howard Zinn and Noam Chomsky—converse with Sasha Lilley about their lives and political philosophies, looking back at eight decades of struggle and theoretical debate. Howard Zinn, interviewed shortly before his death, reflects on the genesis of his politics, from the Civil Rights and anti-Vietnam war movements to opposing empire today, as well as history, art and activism. Noam Chomsky discusses the evolution of his libertarian socialist ideals since childhood, his vision for a future postcapitalist society, and his views on the state, science, the Enlightenment, and the future of the planet.

Noam Chomsky is one of the world's leading intellectuals, the father of modern linguistics, and an outspoken media and foreign policy critic. He is Institute Professor emeritus of linguistics at MIT and the author of numerous books and DVDs including *Hegemony and Survival: America's Quest for Global Dominance, Chomsky on Anarchism, The Essential Chomsky,* and *Crisis and Hope: Theirs and Ours* published by PM Press

Howard Zinn was one of the country's most beloved and respected historians, the author of numerous books and plays including *Marx in Soho, You Can't Be Neutral on a Moving Train*, and the best-selling *A People's History of the United States*, and a passionate activist for radical change.

Sasha Lilley (Interviewer) is a writer and radio broadcaster. She is the co-founder and host of the critically acclaimed program of radical ideas, *Against the Grain*. As program director of KPFA Radio, the flagship station of the Pacifica Network, she headed up such award-winning national broadcasts as *Winter Soldier: Iraq and Afghanistan*. Sasha Lilley is the series editor of PM Press' political economy imprint, Spectre.

"Chomsky is a global phenomenon... perhaps the most widely read voice on foreign policy on the planet."
—*The New York Times Book Review*

The Left Left Behind

Terry Bisson

ISBN: 978-1-60486-086-3
$12.00   128 pages

Hugo and Nebula award-winner Terry Bisson is best known for his short stories, which range from the southern sweetness of "Bears Discover Fire" to the alienated aliens of "They're Made Out of Meat." He is also a 1960's New Left vet with a history of activism and an intact (if battered) radical ideology.

The *Left Behind* novels (about the so-called "Rapture" in which all the born-agains ascend straight to heaven) are among the bestselling Christian books in the US, describing in lurid detail the adventures of those "left behind" to battle the Anti-Christ. Put Bisson and the born-agains together, and what do you get? *The Left Left Behind*—a sardonic, merciless, tasteless, take-no-prisoners satire of the entire apocalyptic enterprise that spares no one—predatory preachers, goth lingerie, Pacifica radio, Indian casinos, gangsta rap, and even "art cars" at Burning Man.

Plus: "Special Relativity," a one-act drama that answers the question: When Albert Einstein, Paul Robeson, and J. Edgar Hoover are raised from the dead at an anti-Bush rally, which one wears the dress? As with all Outspoken Author books, there is a deep interview and autobiography: at length, in-depth, no-holds-barred and all-bets off: an extended tour through the mind and work, the history and politics of our Outspoken Author. Surprises are promised.

"Bisson is a national treasure!"
— John Crowley, author of *Little Big*

"Bisson can charm your toes off!"
— *The Washington Post*

"Bisson's prose is a wonder of seemingly effortless control and precision; he is one of science fiction's most promising short story practitioners, proving that in the genre, the short story remains a powerful, viable and evocative form."
— Reed Business Information, Inc.

Wobblies and Zapatistas: Conversations on Anarchism, Marxism and Radical History

Staughton Lynd and
Andrej Grubačić

ISBN: 978-1-60486-041-2
$20.00 300 pages

Wobblies and Zapatistas offers the reader an encounter between two generations and two traditions. Andrej Grubačić is an anarchist from the Balkans. Staughton Lynd is a lifelong pacifist, influenced by Marxism. They meet in dialogue in an effort to bring together the anarchist and Marxist traditions, to discuss the writing of history by those who make it, and to remind us of the idea that "my country is the world." Encompassing a Left libertarian perspective and an emphatically activist standpoint, these conversations are meant to be read in the clubs and affinity groups of the new Movement.

The authors accompany us on a journey through modern revolutions, direct actions, anti-globalist counter summits, Freedom Schools, Zapatista cooperatives, Haymarket and Petrograd, Hanoi and Belgrade, "intentional" communities, wildcat strikes, early Protestant communities, Native American democratic practices, the Workers' Solidarity Club of Youngstown, occupied factories, self-organized councils and soviets, the lives of forgotten revolutionaries, Quaker meetings, antiwar movements, and prison rebellions. Neglected and forgotten moments of interracial self-activity are brought to light. The book invites the attention of readers who believe that a better world, on the other side of capitalism and state bureaucracy, may indeed be possible.

"There's no doubt that we've lost much of our history. It's also very clear that those in power in this country like it that way. Here's a book that shows us why. It demonstrates not only that another world is possible, but that it already exists, has existed, and shows an endless potential to burst through the artificial walls and divisions that currently imprison us. An exquisite contribution to the literature of human freedom, and coming not a moment too soon."
—David Graeber, author of *Fragments of an Anarchist Anthropology* and *Direct Action: An Ethnography*

Black Flags and Windmills: Hope, Anarchy and the Common Ground Collective

scott crow
with a foreword by
Kathleen Cleaver

ISBN: 978-1-60486-077-1
$20.00 256 pages

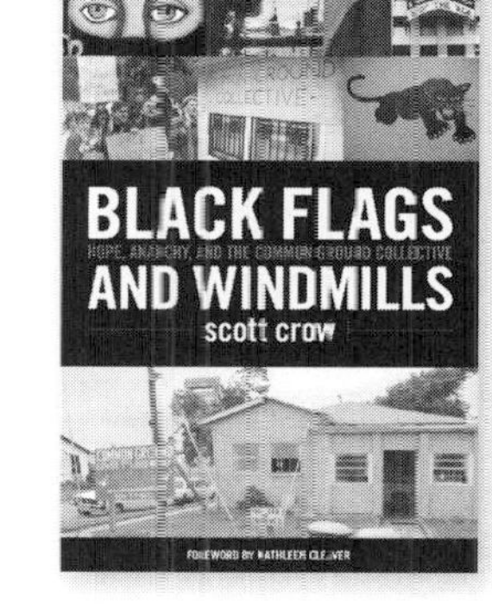

When both levees and governments failed in New Orleans in the Fall of 2005, scott crow headed into the political storm, co-founding a relief effort called the Common Ground Collective. In the absence of local government, FEMA, and the Red Cross, this unusual volunteer organization, based on 'solidarity not charity,' built medical clinics, set up food and water distribution, and created community gardens. They also resisted home demolitions, white militias, police brutality and FEMA incompetence side by side with the people of New Orleans.

crow's vivid memoir maps the intertwining of his radical experience and ideas with Katrina's reality, and community efforts to translate ideals into action. It is a story of resisting indifference, rebuilding hope amidst collapse, and struggling against the grain. *Black Flags and Windmills* invites and challenges all of us to learn from our histories, and dream of better worlds. And gives us some of the tools to do so.

"(scott crow is a) . . . prominent anarchist community organizer behind a host of organizations including Radical Encuentro Camp, and Treasure City Thrift . . ."
— *Austin Chronicle*

". . . a living legend amongst anarchist circles . . ."
— *This American Life*

". . . depending on your sense of humor or your sense of irony . . . (crow and Common Ground) . . . are the good anarchists."
—CNN

". . . crow is a puppetmaster . . ."
— Federal Bureau of Investigation

All Power to the Councils!: A Documentary History of the German Revolution of 1918–1919

Edited and translated by Gabriel Kuhn

ISBN: 978-1-60486-111-2
$26.95 344 pages

The defeat in World War I and the subsequent end of the Kaiserreich threw Germany into turmoil. While the Social Democrats grabbed power, radicals across the country rallied to establish a socialist society under the slogan "All Power to the Councils!" The Spartacus League staged an uprising in Berlin, council republics were proclaimed in Bremen and Bavaria, and workers' revolts shook numerous German towns. The rebellions were crushed by the Social Democratic government with the help of right-wing militias like the notorious Free Corps. This paved the way to a dysfunctional Weimar Republic that witnessed the rise of the National Socialist movement.

The documentary history presented here collects manifestos, speeches, articles, and letters from the German Revolution, introduced and annotated by the editor. Many documents, like the anarchist Erich Mühsam's comprehensive account of the Bavarian Council Republic, are made available in English for the first time. The volume also includes appendixes portraying the Red Ruhr Army that repelled the reactionary Kapp Putsch in 1920, and the communist bandits that roamed Eastern Germany until 1921. *All Power to the Councils!* provides a dynamic and vivid picture of a time with long-lasting effects for world history. A time that was both encouraging and tragic.

"The councils of the early 20th century, as they are presented in this volume, were autonomous organs of the working class beyond the traditional parties and unions. They had stepped out of the hidden world of small political groups and represented a mass movement fighting for an all-encompassing council system."
— Teo Panther, editor of *Alle Macht den Räten: Novemberrevolution 1918*

"The German Revolution of 1918–1919 and the following years mark an exceptional period in German history. This collection brings the radical aspirations of the time alive and contains many important lessons for contemporary scholars and activists alike."
— Markus Bauer, Free Workers' Union, FAU-IAA

West of Eden: Communes and Utopia in Northern California

Edited by Iain Boal, Janferie Stone, Michael Watts, and Cal Winslow

ISBN: 978-1-60486-427-4
$24.95 304 pages

In the shadow of the Vietnam War, a significant part of an entire generation refused their assigned roles in the American century. Some took their revolutionary politics to the streets, others decided simply to turn away, seeking to build another world together, outside the state and the market. *West of Eden* charts the remarkable flowering of communalism in the '60s and '70s, fueled by a radical rejection of the Cold War corporate deal, utopian visions of a peaceful green planet, the new technologies of sound and light, and the ancient arts of ecstatic release.

Using memoir and flashbacks, oral history and archival sources, *West of Eden* explores the deep historical roots and the enduring, though often disavowed, legacies of the extraordinary pulse of radical energies that generated forms of collective life beyond the nuclear family and the world of private consumption, including the contradictions evident in such figures as the guru/predator or the hippie/entrepreneur. There are vivid portraits of life on the rural communes of Mendocino and Sonoma, and essays on the Black Panther communal households in Oakland, the latter-day Diggers of San Francisco, the Native American occupation of Alcatraz, the pioneers of live/work space for artists, and the Bucky dome as the iconic architectural form of the sixties.

West of Eden is not only a necessary act of reclamation, helping to record the unwritten stories of the motley generation of communards and antinomians now passing, but is also intended as an offering to the coming generation who will find here, in the rubble of the twentieth century, a past they can use—indeed one they will need—in the passage from the privations of commodity capitalism to an ample life in common.

"As a gray army of undertakers gather in Sacramento to bury California's great dreams of equality and justice, this wonderful book, with its faith in the continuity of our state's radical-communitarian ethic, replants the seedbeds of defiant imagination and hopeful resistance."
—Mike Davis, author of *City of Quartz* and *Magical Urbanism*

Ned Ludd & Queen Mab: Machine-Breaking, Romanticism, and the Several Commons of 1811–12

Peter Linebaugh

ISBN: 978-1-60486-704-6
$6.95 48 pages

Peter Linebaugh, in an extraordinary historical and literary *tour de force*, enlists the anonymous and scorned 19th century loom-breakers of the English midlands into the front ranks of an international, polyglot, many-colored crew of commoners resisting dispossession in the dawn of capitalist modernity.

"Sneering at the Luddites is still the order of the day. Peter Linebaugh's great act of historical imagination stops the scoffers in their tracks. It takes the cliche of 'globalization' and makes it live: the Yorkshire machine-breakers are put right back in the violent world economy of 1811–12, in touch with the Atlantic slave trade, Mediterranean agri-business, the Tecumseh rebellion, the brutal racism of London dockland. The local and the global are once again shown to be inseparable—as they are, at present, for the machine-breakers of the new world crisis."
— T.J. Clark, author of *The Absolute Bourgeois* and *Image of the People*

"My benediction"
— E.J. Hobsbawm, author of *Primitive Rebels* and *Captain Swing*

"E.P. Thompson, you may rest now. Linebaugh restores the dignity of the despised luddites with a poetic grace worthy of the master. By a stunning piece of re-casting we see them here not as rebels against the future but among the avant-garde of a planetary resistance movement against capitalist enclosures in the long struggle for a different future. Byron, Shelley, listen up! Peter Linebaugh's Ned Ludd and Queen Mab *does for 'technology' what his* London Hanged *did for 'crime.' Where was I that day in Bloomsbury when he delivered this commonist manifesto for the 21st century? The Retort Pamphet series is off to a brilliant start."*
— Mike Davis, author of *Planet of Slums* and *Buda's Wagon*